OFFICIAL SQA PAST PAPERS WITH ANSWERS

STANDARD GRADE | GENERAL | CREDIT

GEOGRAPHY
2006-2010

© Scottish Qualifications Authority

First exam published in 2006.
Published by Bright Red Publishing Ltd, 6 Stafford Street, Edinburgh EH3 7AU
tel: 0131 220 5804 fax: 0131 220 6710 info@brightredpublishing.co.uk www.brightredpublishing.co.uk

ISBN 978-1-84948-093-2

A CIP Catalogue record for this book is available from the British Library.

Bright Red Publishing is grateful to the copyright holders, as credited on the final page of the book, for permission to use their material.
Every effort has been made to trace the copyright holders and to obtain their permission for the use of copyright material.
Bright Red Publishing will be happy to receive information allowing us to rectify any error or omission in future editions.

STANDARD GRADE | GENERAL

2006

[BLANK PAGE]

FOR OFFICIAL USE

KU	ES

Total Marks

1260/403

NATIONAL
QUALIFICATIONS
2006

WEDNESDAY, 10 MAY
10.25 AM–11.50 AM

GEOGRAPHY
STANDARD GRADE
General Level

Fill in these boxes and read what is printed below.

Full name of centre

Town

Forename(s)

Surname

Date of birth

Day Month Year   Scottish candidate number   Number of seat

1 Read the whole of each question carefully before you answer it.

2 Write in the spaces provided.

3 Where boxes like this ☐ are provided, put a tick ✓ in the box beside the answer you think is correct.

4 Try all the questions.

5 Do not give up the first time you get stuck: you may be able to answer later questions.

6 Extra paper may be obtained from the invigilator, if required.

7 Before leaving the examination room you must give this book to the invigilator. If you do not, you may lose all the marks for this paper.

SCOTTISH
QUALIFICATIONS
AUTHORITY

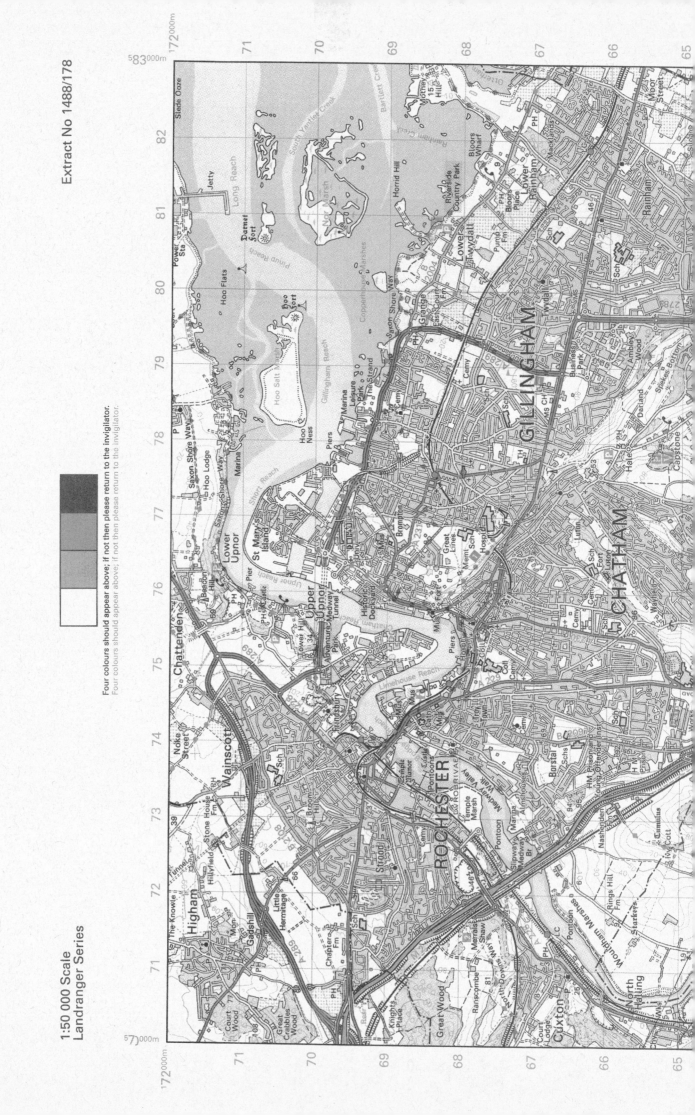

1:50 000 Scale
Landranger Series

Extract No 1488/178

Four colours should appear above; if not then please return to the invigilator.
Four colours should appear above; if not then please return to the invigilator.

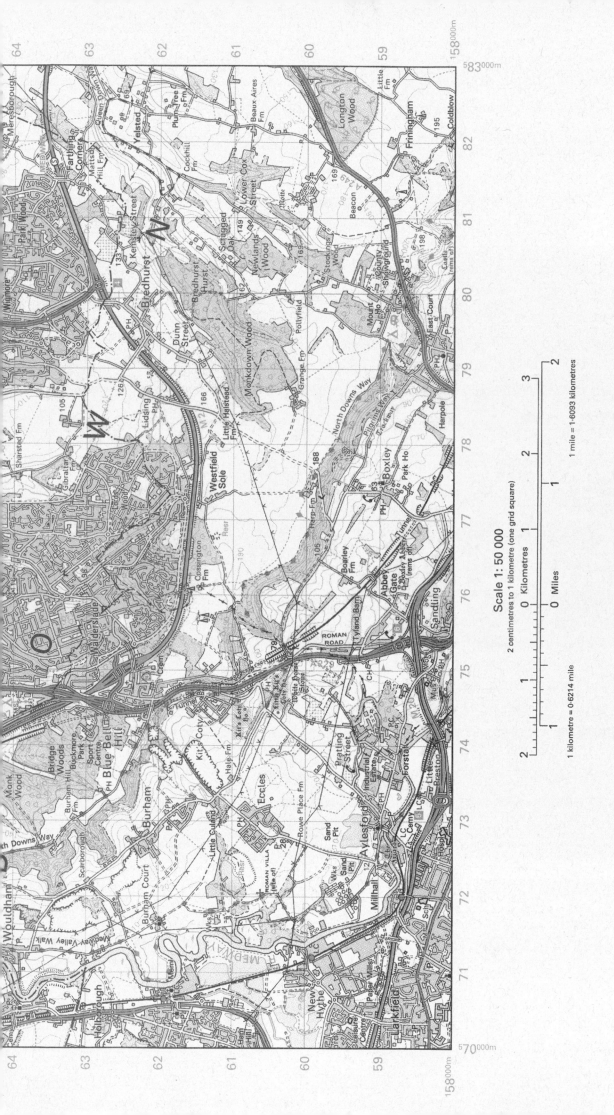

Scale 1: 50 000

2 centimetres to 1 kilometre (one grid square)

Kilometres

Miles

1 kilometre = 0·6214 mile

1 mile = 1·6093 kilometres

1.

Reference Diagram Q1

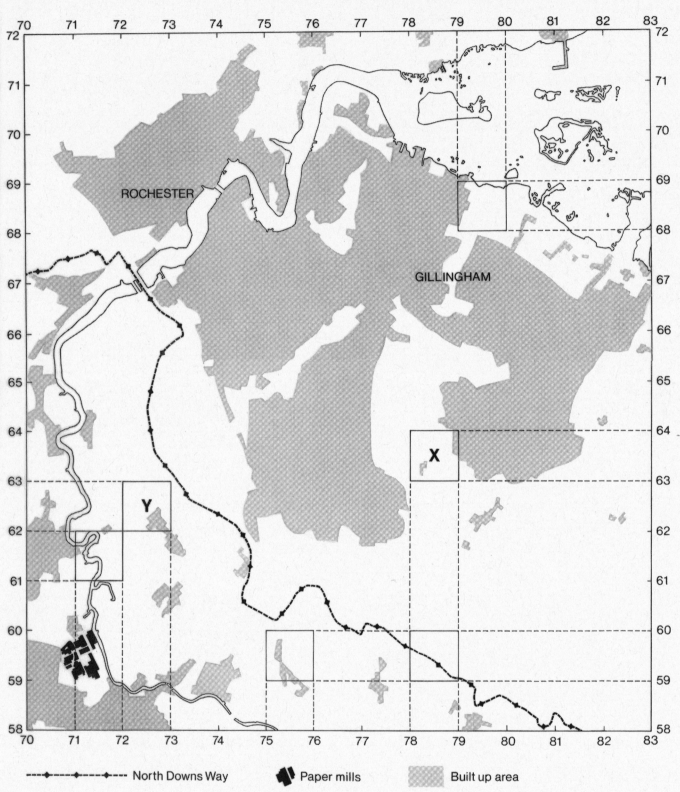

KU | ES

Marks

1. (continued)

Look at the Ordnance Survey Map Extract (No 1488/178) of the Thames Estuary area and Reference Diagram Q1 on Page two.

(*a*)   **Explain** why an oxbow lake is likely to develop in grid square 7161.

You may use diagrams to illustrate your answer.

(7161)
The meanders is very closer so by hydraulic action and abrasion the water will erode passed the corner through the river then

the water no longer flows through leaving it to dry up

4

(*b*)   Find the North Downs Way on Reference Diagram Q1 and on the Ordnance Survey Map.

The North Downs Way is a footpath for recreational walkers.

Using map evidence, give the advantages **and** disadvantages of this route.

4

[Turn over

Marks

1. **(continued)**

(*c*) Give different reasons why there are trees in the following grid squares.

7559 _____

_____

7859 _____

_____

7968 _____

_____   **3**

(*d*) It is proposed to develop either Area X (7863) or Area Y (7262) for new housing.

Which area, X or Y, do you think is more suitable?

Explain your answer **in detail**.

_____

_____

_____

_____

_____

_____   **4**

Marks

1. (continued)

(*e*) There is a large paper mill in grid square 7159.

Explain why this site is a suitable location for the paper mill.

You **must** use map evidence.

_____

_____

_____

_____

_____

_____

4

(*f*) Rochester and Gillingham are built either side of the River Medway.

In what ways has the River Medway **both** benefited **and** created problems for these settlements?

_____

_____

_____

_____

_____

4

[Turn over

Marks

2.   **Reference Diagram Q2: A Landscape of Glacial Deposition**

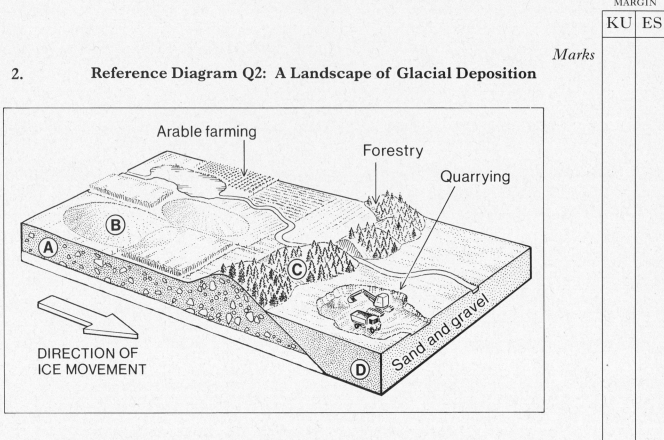

(*a*)   Match each of the features of glacial deposition in the table to the correct letter (A, B, C, D) on the Reference Diagram.

Feature	Letter
Drumlin	
Terminal moraine	
Outwash plain	
Boulder clay	

3

Marks

2. (continued)

(b) **Explain** why **two** of the land uses shown on Reference Diagram Q2
 are suitable for the areas in which they are located.

Land Use 1 _____

_____

_____

_____

_____

Land Use 2 _____

_____

_____

_____

_____

4

[Turn over

Marks

3.   **Reference Diagram Q3A: Janice's Radio Phone-in**

KU | ES

Marks

3. (continued)

Reference Diagram Q3B: Synoptic Chart, 10 August 2004

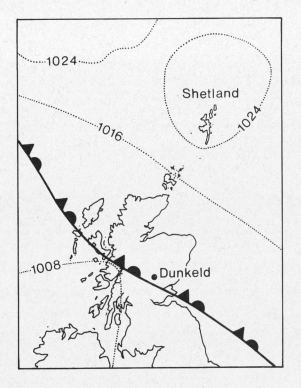

Look at Reference Diagrams Q3A and Q3B.

Explain the different weather experiences which Janice's two callers had on 10 August 2004.

_____

_____

_____

_____

_____

4

[Turn over

Marks KU ES

4.   **Reference Diagram Q4A: Data for six Weather Stations in Canadian Tundra on 3 May 2004**

Station	Hours of Sunshine	Mean Temperature °C
A	2	1
B	6	4
C	8	6
D	3	2
E	5	3
F	2	4

Reference Diagram Q4B: Scattergraph

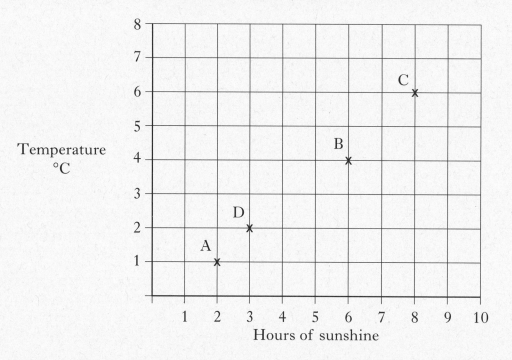

(a) Use the information in Reference Diagram Q4A to complete the scattergraph in Reference Diagram Q4B above.   2

(b) "The scattergraph shows that temperature in the Tundra is directly linked to the hours of sunshine."

Do you agree with this statement?

Give reasons for your answer.

_____

_____

_____

_____   2

KU | ES

Marks

5. **Reference Diagram Q5: Developments in the North Sea**

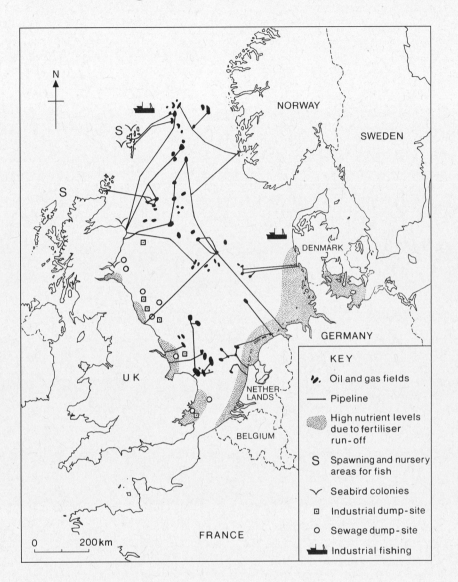

> "We must stop exploiting the North Sea. We are destroying its environment."
>
> EU Politician

Do you agree with the statement above?

Give reasons for your answer.

_____

_____

_____

_____

_____

_____

4

6. **Reference Diagram Q6A:**
    Newspaper Headlines

    Reference Diagram Q6B: Examples of
    Diversification* on a Farm

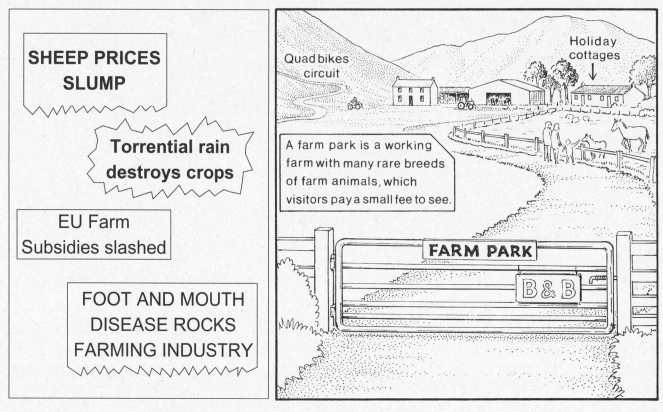

SHEEP PRICES SLUMP

Torrential rain destroys crops

EU Farm Subsidies slashed

FOOT AND MOUTH DISEASE ROCKS FARMING INDUSTRY

Quad bikes circuit

Holiday cottages

A farm park is a working farm with many rare breeds of farm animals, which visitors pay a small fee to see.

FARM PARK

B & B

*Diversification = using other land uses to improve farm income

Explain, in some detail, how diversification can help farmers overcome the problems shown in Reference Diagram Q6A.

KU | ES

Marks

4

KU | ES

Marks

7.   **Reference Diagram Q7: Factors affecting Location of Industry**

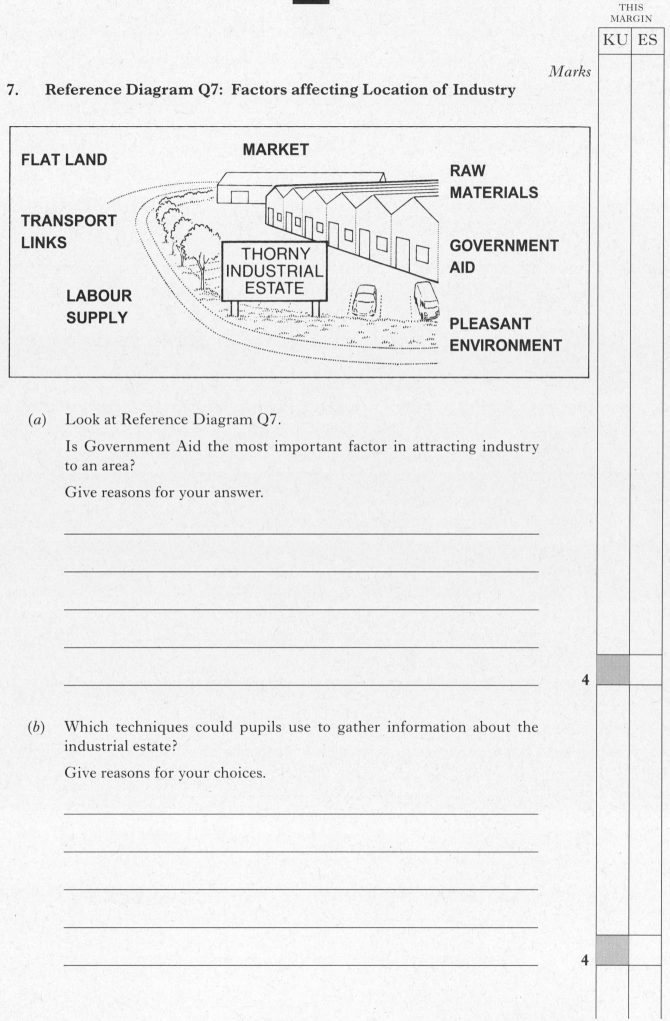

FLAT LAND

MARKET

RAW MATERIALS

TRANSPORT LINKS

THORNY INDUSTRIAL ESTATE

GOVERNMENT AID

LABOUR SUPPLY

PLEASANT ENVIRONMENT

(a)   Look at Reference Diagram Q7.

Is Government Aid the most important factor in attracting industry to an area?

Give reasons for your answer.

4

(b)   Which techniques could pupils use to gather information about the industrial estate?

Give reasons for your choices.

4

Marks

8.   **Reference Diagram Q8: Population Data for two Countries**

Population Data A   **Population Data B**

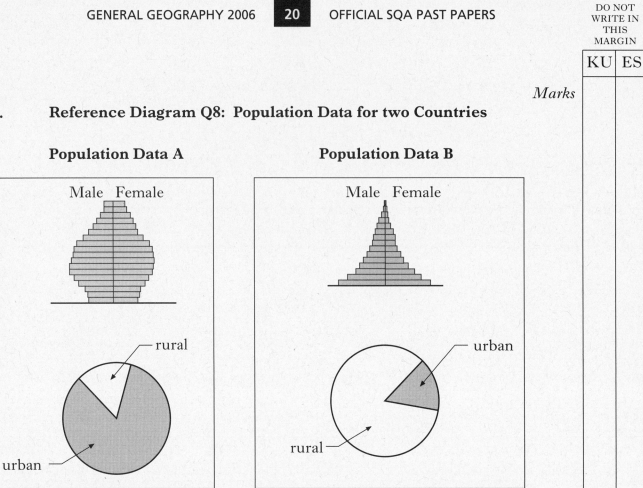

Which set of population data, A or B, is more typical of an Economically Less Developed Country (ELDC)?

Give reasons for your choice.

_____

_____

_____

_____

_____

_____

4

9. **Reference Diagram Q9: Japan's Trade Pattern**

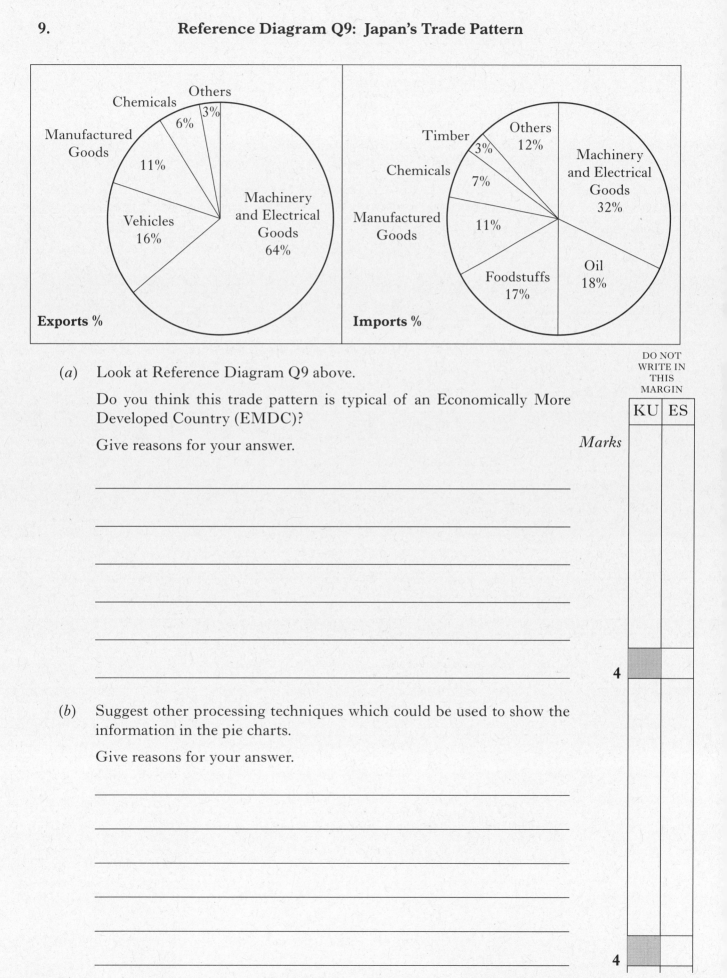

Exports %

Imports %

(a) Look at Reference Diagram Q9 above.

Do you think this trade pattern is typical of an Economically More Developed Country (EMDC)?

Give reasons for your answer.

Marks

4

(b) Suggest other processing techniques which could be used to show the information in the pie charts.

Give reasons for your answer.

4

Marks

10.   **Reference Diagram Q10: Uses of Aid**

Better Farming Techniques
- Increased mechanisation
- Irrigation
- More fertiliser

Education Opportunities
- More teachers
- More schools and colleges

← **AID** →

Improved Water Supply
- Water control project
- Wells
- Dams

Transport and Communications improved
- Better roads
- Better telecommunications

Which **two** of the above uses of aid do you think would be of **most** benefit to Economically Less Developed Countries (ELDCs)?

Give reasons for your choices.

Choice 1 _____

_____

_____

_____

_____

Choice 2 _____

_____

_____

_____

_____

4

[END OF QUESTION PAPER]

STANDARD GRADE | CREDIT

2006

[BLANK PAGE]

C

1260/405

| NATIONAL QUALIFICATIONS 2006 | WEDNESDAY, 10 MAY 1.00 PM – 3.00 PM | GEOGRAPHY STANDARD GRADE Credit Level |

All questions should be attempted.

Candidates should read the questions carefully. Answers should be clearly expressed and relevant.

Credit will always be given for appropriate sketch-maps and diagrams.

Write legibly and neatly, and leave a space of about one cm between the lines.

All maps and diagrams in this paper have been printed in black only: no other colours have been used.

SCOTTISH
QUALIFICATIONS
AUTHORITY

©

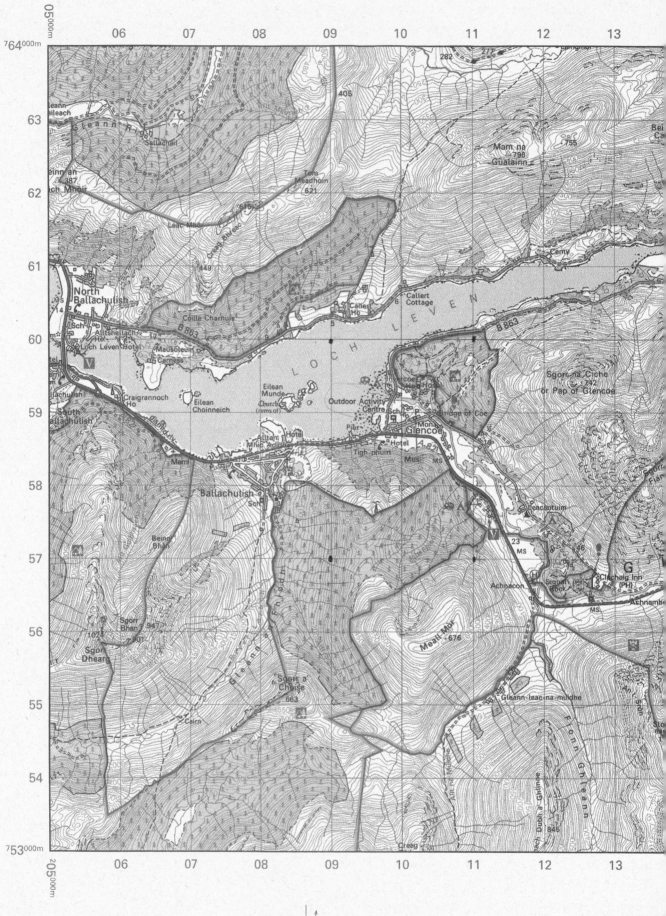

Four colours should appear abov
Four colours should appear abov

Grid North
True North
Magnetic North
Diagrammatic only

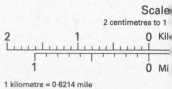

Scale
2 centimetres to 1

2   1   0   Kil

1   0   Mi

1 kilometre = 0·6214 mile

23000m

764000m

63

62

61

60

59

58

57

56

55

54

753000m

15   16   17   18   19   20   21   22

Fhuarain

Mamore Lodge (Hotel)

Meall an Doire Dharaich

Meall na Duibhe
673

Kinlochmore

Waterfall

Kinlochleven

Waterfall

64   Waterfall

734

Garbh Bheinn
867

Meall Roigh a' Bhricleathaid

508

Sron a' Choire Odhair Bhig

Old Military Road

Meall Garbh

866

Stob Coire Leith

Meall Dearg
953

Aonach Eagach

The Chancellor

Am Bodach

873
Sron Gharbh

943

903

Beinn Bheag
616

Stob Mhic Mhartuin
707

Devil's Staircase

A82

Achtriochtan

PASS OF GLENCOE

A Chailleach

Allt-na-reigh

The Study

MS   Cairn

MS

Altnafeadh

River Coe

Loch Achtriochtan

Ossian's Cave

Waterfall
Meeting of Three Waters

Waterfall

Lochan na Fola

Lagangarbh

A82

The Three Sisters

Aonach Dubh
892

Gearr Aonach
692

Stob nan Cabar

River Coupall

Stob Coire nan Lochan
1115

811

Lairig Eilde

Buachaille Etive Beag

Stob Coire Raineach
925

Coire na Tulaich

902

Stob Dearg
1022

1150

931

Lairig Gartain

902

Buachaille Etive Mor

Coire Cloiche Finne

Stob Coire Sgreamhach
1070

Stob Dubh
958

ROYAL   FOREST

1011
Stob na Doire

n Bian

941

15   16   17   18   19   20   21   22

23000m

Extract produced by Ordnance Survey 2005
© Crown copyright 2002. All rights reserved.

1.

Reference Diagram Q1A

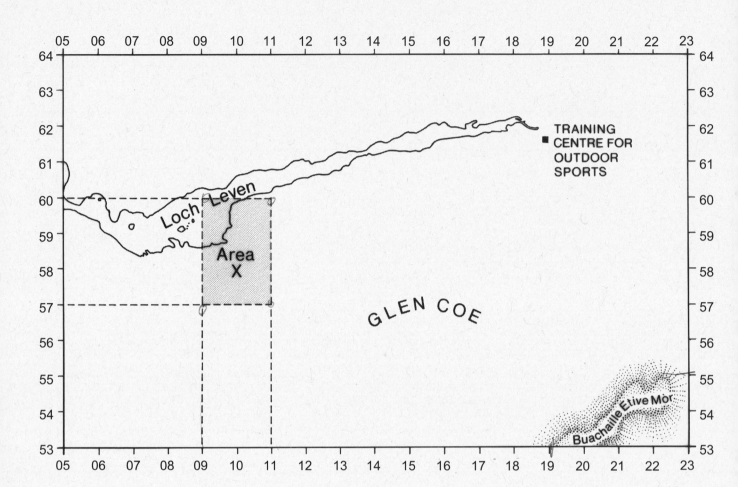

Marks

KU | ES

1. (continued)

This question refers to the OS Map Extract (No 1489/41) of the Glen Coe area and Reference Diagram Q1A on *Page two*.

Reference Diagram Q1B: View looking West from 197618

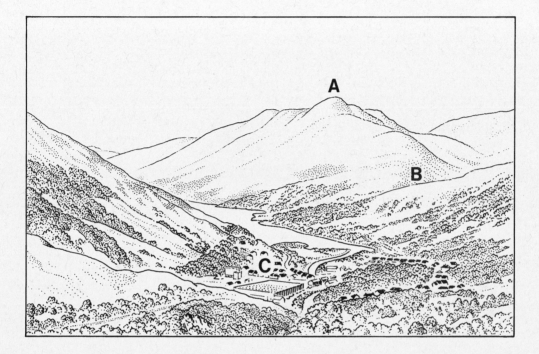

(a)   Look at Reference Diagram Q1B and the map extract.

Reference Diagram Q1B is a view looking west from 197618.

Identify the **three** features marked A, B and C.

Choose from:

Allt Nathrach;   Kinlochleven;   Mam na Gualainn;   Kinlochmore;

Beinn na Caillich.   3

(b)   (i)   Match each of the features named below with the correct grid reference.

Features: **arete;   hanging valley;   truncated spur;   corrie**.

Choose from grid references: 165553,   057563,   197584,   201556.   3

(ii)   **Explain** how **one** of the features listed in (b)(i) was formed.

You may use diagrams to illustrate your answer.   4

[Turn over

Marks

KU F

1. (continued)

(*c*) | "Glen Coe and the surrounding area is one of Scotland's most popular tourist areas." |

Spokesperson for the Scottish Tourist Industry

Part of the disused aluminium works at Kinlochleven (1861, 1862) has been converted into a training centre for outdoor sports such as climbing, walking and ice climbing.

Using map evidence to support your answer, state whether or not you think this is a good location for an outdoor sports centre.

(*d*)   Look at Reference Diagram Q1A.

Find Area X on the OS map extract.

Give reasons for the different land uses in this area.

5

Reference Diagram Q1C: A Wind Farm

(*e*)   A developer is proposing to build a wind farm of up to thirty wind turbines on Buachaille Etive Mòr (see Reference Diagrams Q1A and Q1C).

Do you think this proposal should go ahead?

You **must** use map evidence to support your answer.

2.   **Reference Diagram Q2: Synoptic Chart for 15 January 1995**

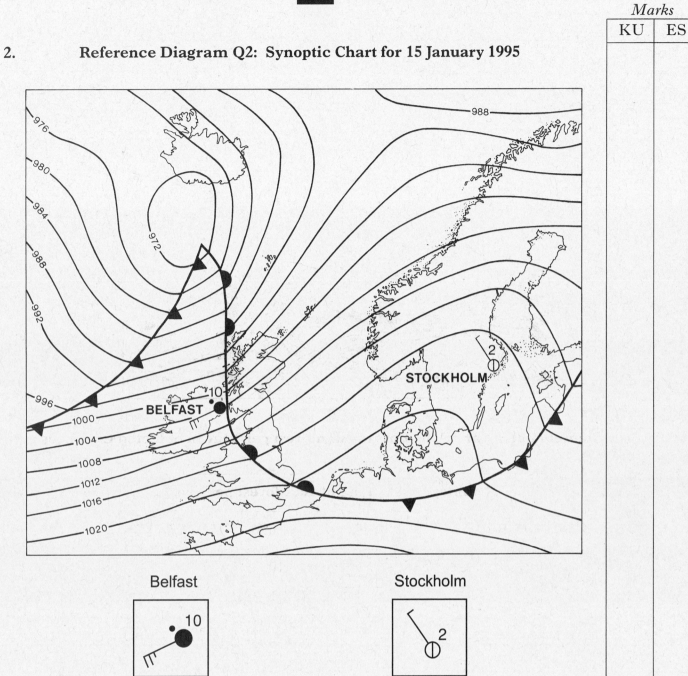

Belfast   Stockholm

Look at Reference Diagram Q2.

Explain the **differences** in the weather conditions between Belfast and Stockholm.

6

[Turn over

3.

Reference Diagram Q3A: Destruction of Rainforest

Reference Diagram Q3B: Loss of Rainforest per Year in Selected Countries

Country	Loss of Rainforest per Year
Brazil	6%
Indonesia	10%
Venezuela	12%

Look at Reference Diagrams Q3A and Q3B.

Explain why the world's rainforests continue to be destroyed.

5

Marks

KU | ES

4.   **Reference Diagram Q4A: Recent Changes in Farming**

> • Farmers paid to "set aside" land
> • Farmers converting to produce organic food
> • Removal of hedges to make fields larger
> • Increased mechanisation

(a) Study Reference Diagram Q4A.

Do you think these changes **benefit** the countryside?

Give reasons for your answer.

6

Reference Diagram Q4B: Land Use Data for Crow Farm East Anglia

Land Use	Area (hectares)
Barley	13·5
Wheat	12·5
Farm yard and buildings	1·2
Sugar beet	12·0
Vegetables	5·5
Set aside	8·5

(b) Study Reference Diagram Q4B.

Give **two** other techniques which would be appropriate to process this data.

Explain your choices.

5

[Turn over

Marks

KU | E

5.

Reference Diagram Q5A: Land Values in a City

Land
Values
£

CBD | Inner city | Suburbs | Edge of city

Reference Diagram Q5B: Features of Housing Areas

Inner City	Suburbs
Nearer CBD	**Nearer edge of city**
In grid-iron pattern with long, straight streets	Varied street pattern with many short streets in a cul-de-sac arrangement
Houses close to industry	Housing separate from industry
Tenements and/or terraces	Variety of house types, with many detached and semi-detached
Little open space/greenery	More spacious with many gardens
Environmental problems	Pleasant environment

KU	ES

5. (continued)

Look at Reference Diagrams Q5A and Q5B.

(a) **Explain** the features of **either** the Inner City **or** the Suburbs shown in Reference Diagram Q5B.

(b) Describe techniques which could be used to gather information about differences between the environments of two urban areas.

Give reasons for your choice of techniques.

[Turn over

6.   **Reference Diagram Q6: Population Growth in Tokyo and Jakarta**

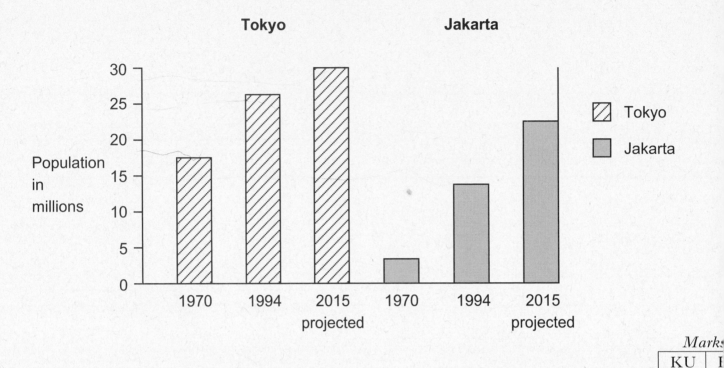

(a) Compare **in detail** the population growth in the two cities.

Tokyo is the capital of Japan, an Economically More Developed Country (EMDC).

Jakarta is the capital of Indonesia, an Economically Less Developed Country (ELDC).

(b) This population growth will cause more problems for Tokyo than for Jakarta.

Do you agree?

Give reasons for your answer.

Marks

KU I

7.   **Reference Diagram Q7: Factors affecting Population Distribution**

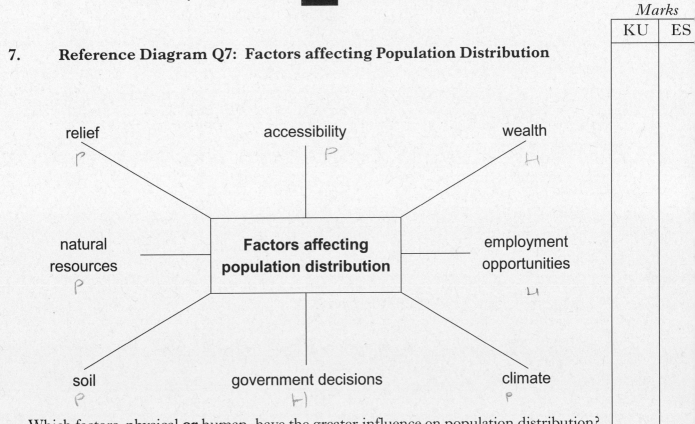

Which factors, physical **or** human, have the greater influence on population distribution?

Give reasons for your choice.

6

[Turn over for Question 8 on *Page twelve*

Page eleven

8.   **Reference Diagram Q8: Location of the Three Gorges Dam**

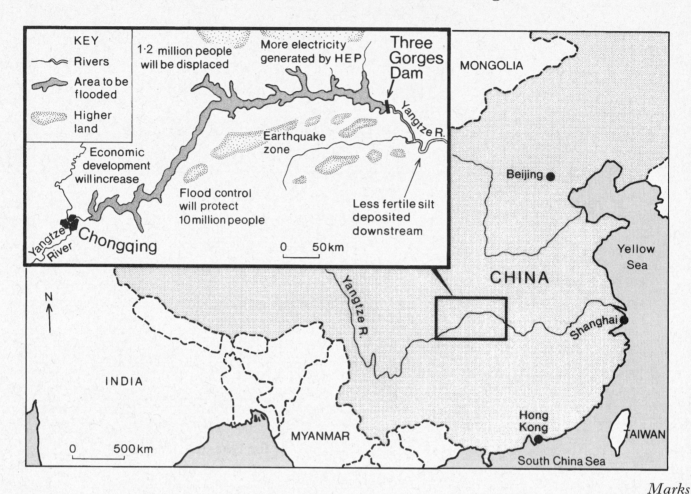

Marks

KU	ES

The Three Gorges Dam project was built with money from China and investments from Japan, Canada, Germany and Switzerland. These investments were made in order to develop trade links with China.

What are the advantages **and** disadvantages for China and its trading partners of using these investments to build the Three Gorges Dam?

6

[END OF QUESTION PAPER]

STANDARD GRADE | GENERAL

2007

[BLANK PAGE]

FOR OFFICIAL USE

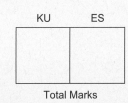

KU   ES

Total Marks

1260/403

NATIONAL
QUALIFICATIONS
2007

TUESDAY, 8 MAY
10.25 AM–11.50 AM

GEOGRAPHY
STANDARD GRADE
General Level

Fill in these boxes and read what is printed below.

Full name of centre

Town

Forename(s)

Surname

Date of birth

Day Month Year   Scottish candidate number   Number of seat

1 Read the whole of each question carefully before you answer it.

2 Write in the spaces provided.

3 Where boxes like this ☐ are provided, put a tick ✓ in the box beside the answer you think is correct.

4 Try all the questions.

5 Do not give up the first time you get stuck: you may be able to answer later questions.

6 Extra paper may be obtained from the invigilator, if required.

7 Before leaving the examination room you must give this book to the invigilator. If you do not, you may lose all the marks for this paper.

Scottish
Qualifications
Authority

1:50 000 Scale
Landranger Series

Four colours should appear abov
Four colours should appear abov

Scale 1: 50 000

2 centimetres to 1 kilometre (one grid square)

1 kilometre = 0·6214 mile

Grid North

Magnetic North

True North

Diagrammatic only

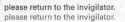

please return to the invigilator.
please return to the invigilator.

3

2

= 1·6093 kilometres

1.

Reference Diagram Q1: The Dorchester Area

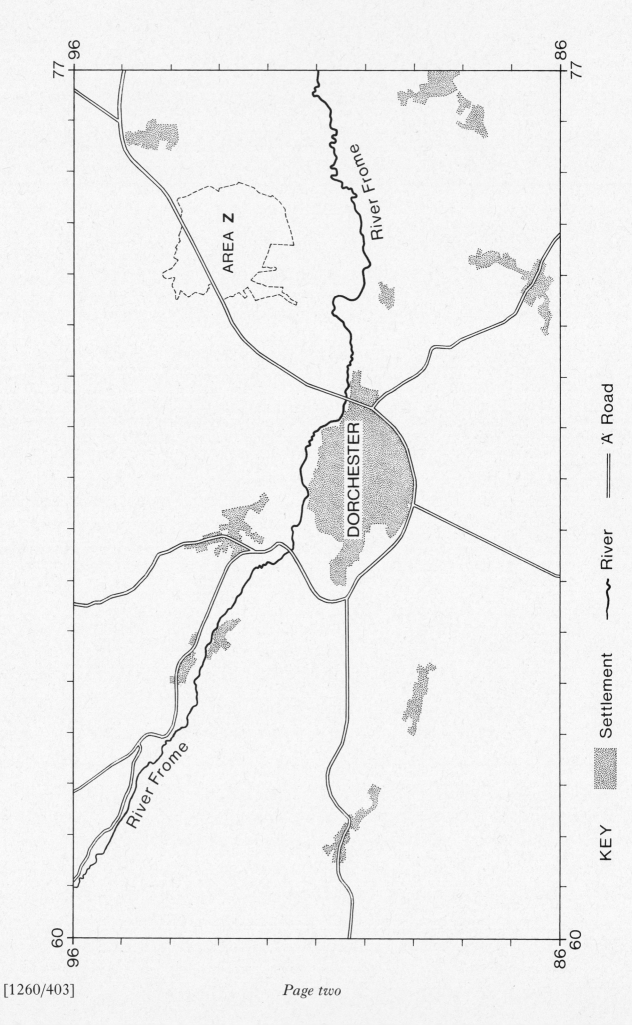

KEY

░░░ Settlement   ～ River   —— 'A' Road

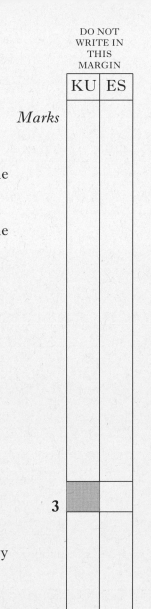

Marks

1. (continued)

Look at the Ordnance Survey Map Extract (No 1557/194) of the Dorchester area and Reference Diagram Q1 on Page two.

(*a*) Complete the table below by matching the physical features to the correct grid references.

Choose from: 6586   7490   6193   6286

Physical Feature	Grid Square
Steep southwest facing slopes	
Flat land	
Broad ridge running East–West	
V-shaped valley	

3

(*b*) Describe the **physical** features of the River Frome **and** its valley between grid references 610959 and 700909.

_____

_____

_____

_____

_____

4

_____

[Turn over

Marks

1. (continued)

(c)   Give **two** techniques which could be used to gather information about the physical characteristics of the River Frome.

Give reasons for your choice of techniques.

Technique 1: _____

Reason: _____

_____

Technique 2: _____

Reason: _____

_____   **4**

(d)   It is proposed to develop Area Z into a country park (see Reference Diagram Q1).

Using map evidence, give arguments for **and** against this proposal.

For: _____

_____

_____

_____

Against: _____

_____

_____

_____   **4**

Marks

1. (continued)

(*e*)   **Explain** why Dorchester has expanded west into grid square 6790 and not north into grid square 6991.

Refer to both grid squares in your answer.

_____

_____

_____

_____

_____

_____

3

(*f*)   Which is the more likely function of Dorchester?

Tick (✓) your choice.

Tourist resort ☐   Market town ☐

Give map evidence to support your choice.

_____

_____

_____

_____

_____

3

(*g*)   What are the **disadvantages** of the location of Maiden Castle Farm (grid square 6789)?

_____

_____

_____

_____

_____

3

Marks

2.   **Reference Diagram Q2: A Glaciated Lowland**

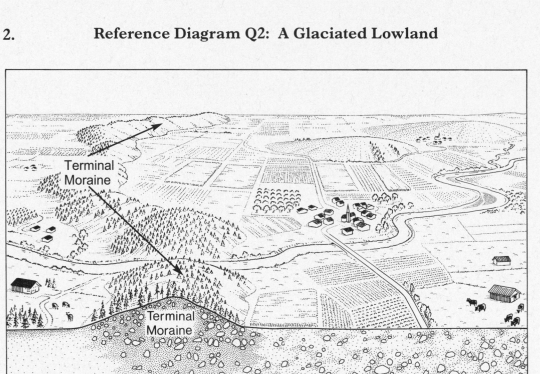

Look at Reference Diagram Q2.

Explain how a terminal moraine is formed.

You may wish to use diagram(s) to illustrate your answer.

_____

_____

_____

_____

(*Space for diagrams*)

3

Marks

3.   Reference Diagram Q3: Air Masses affecting the British Isles

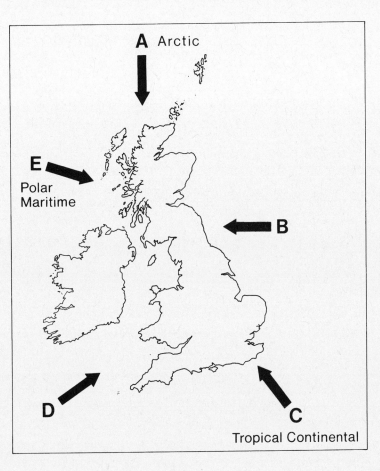

Look at Reference Diagram Q3.

(*a*)   Name air masses B and D.

B _____

D _____

(*b*)   **Describe** the benefits **and** problems of a long spell of weather caused
by air mass C in summer.

_____

_____

_____

_____

_____

[Turn over

Marks   KU   ES

4.   **Reference Diagram Q4A: A Tropical Rainforest Climate Graph**

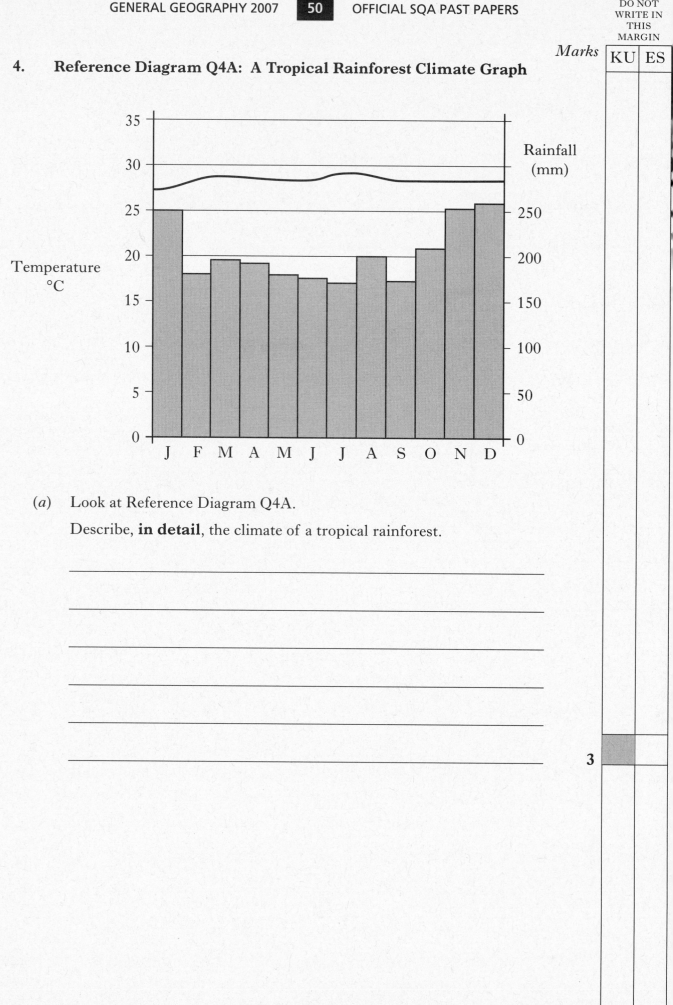

(a)   Look at Reference Diagram Q4A.

Describe, **in detail**, the climate of a tropical rainforest.

_____

_____

_____

_____

_____

_____   **3**

4. (continued)

Reference Diagram Q4B: Developments in the Rainforests of Brazil

(b) Look at Reference Diagram Q4B.

"Developments in rainforests have brought many benefits to local people."

Do you agree with the above statement?

Explain your answer.

_____

_____

_____

_____

_____

_____

_____

4

[Turn over

5.   **Reference Diagram Q5A: Brockan Farm in 1977**

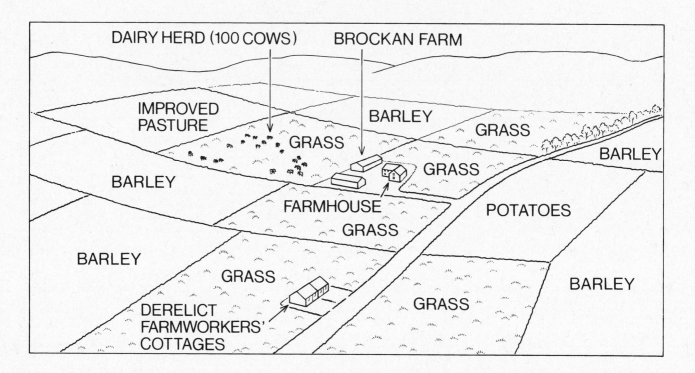

Reference Diagram Q5B: Brockan Farm in 2007

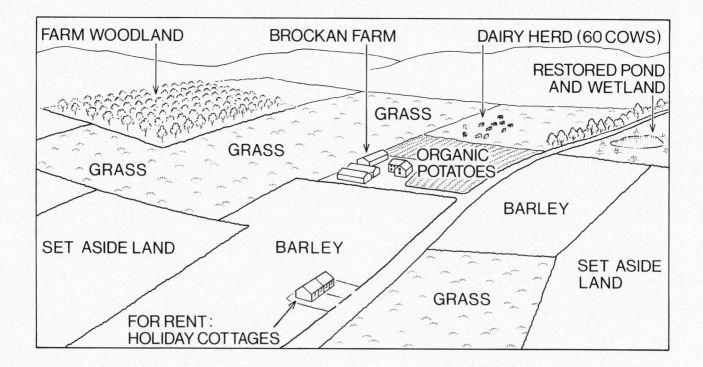

Marks

5. (continued)

(a) Study Reference Diagrams Q5A and Q5B. Suggest reasons for the changes shown on Brockan Farm between 1977 and 2007.

_____

_____

_____

_____

_____

_____

4

Reference Diagram Q5C: Land Use on Brockan Farm, 2007

Land Use	% of Total Land
Grass	45%
Barley	25%
Woodland	10%
Set aside	10%
Organic potatoes	5%
Restored wetland	5%

(b) Study Reference Diagram Q5C.

Complete the divided bar graph below.

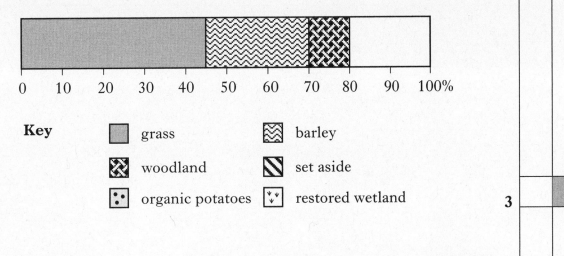

3

[Turn over

6.   **Reference Diagram Q6: Location of a Cement Works**

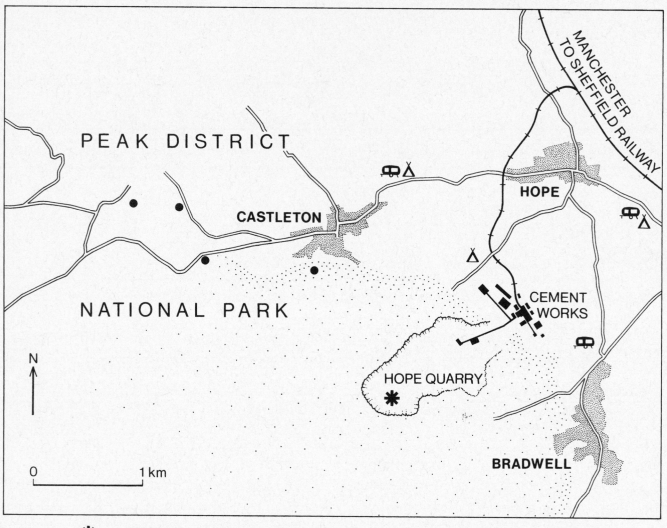

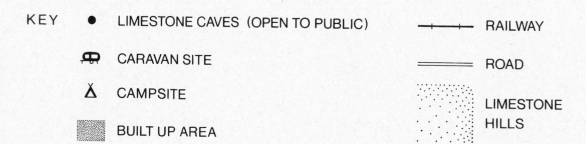

✱ LIMESTONE, THE MAIN RAW MATERIAL IN THE MANUFACTURE OF CEMENT

KEY   ● LIMESTONE CAVES (OPEN TO PUBLIC)   ┼────┼ RAILWAY

        ⊞ CARAVAN SITE   ═════ ROAD

        △ CAMPSITE   ┊┊┊┊ LIMESTONE
                                                         HILLS
        ▨ BUILT UP AREA

Marks

6. (continued)

Study Reference Diagram Q6.

Do you think this is a good location for a cement works?

Give reasons for your answer.

_____

_____

_____

_____

_____

_____

4

[Turn over

Marks

7.   **Reference Diagram Q7: Population Data—India**

Year	Population in Millions
1945	336
1955	395
1965	482
1975	600
1985	749
1995	934
2005	1095

Look at Reference Diagram Q7.

(*a*)   "In an Economically Less Developed Country such as India, the population figures taken from census records are likely to be unreliable."

UN spokesperson

Do you agree with the above statement?

Give reasons for your answer.

_____

_____

_____

_____

_____

_____

_____

3

KU	ES

Marks

7. (continued)

(b) What other techniques could be used to show the information in Reference Diagram Q7?

Explain your choice(s).

_____

_____

_____

_____

_____

_____

4

[Turn over

KU	ES

Marks

8.   **Reference Diagram Q8: Migration from Haiti to USA, 1990–95**

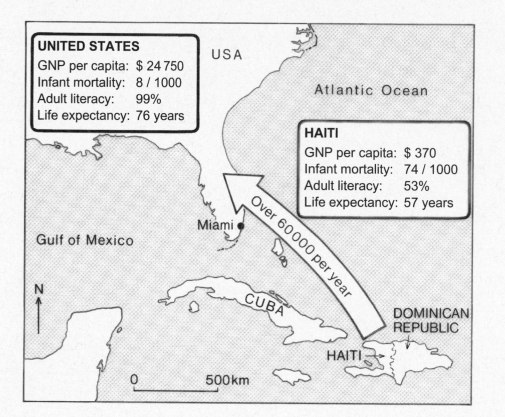

Look at Reference Diagram Q8.

Referring to the data shown, **explain** why so many people migrated from Haiti to USA between 1990 and 1995.

4

9. **Reference Diagram Q9: Selected World Oil Consumption, 2005**

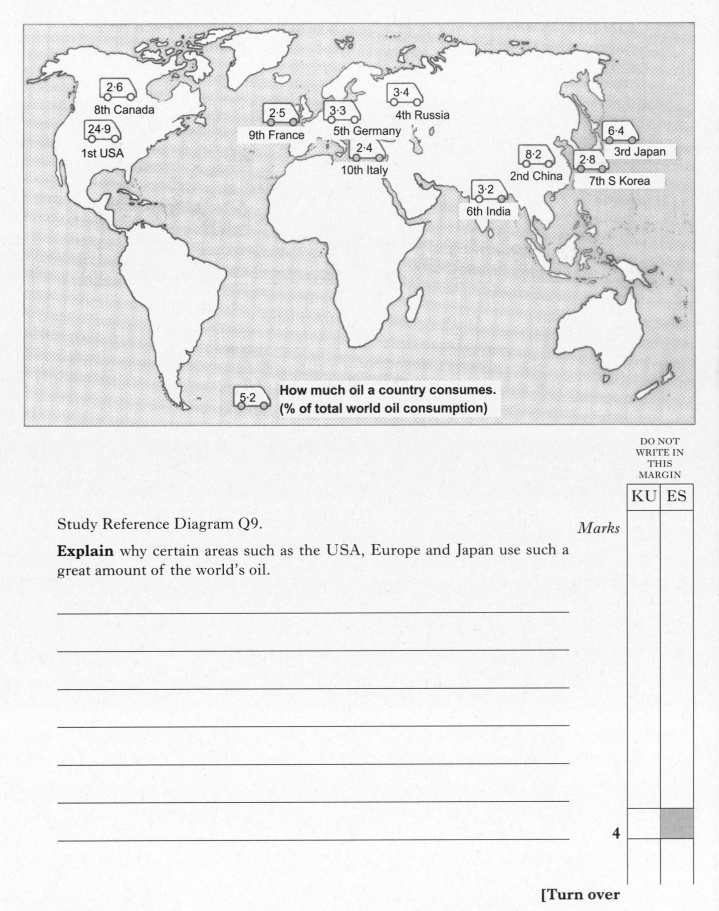

How much oil a country consumes.
(% of total world oil consumption)

Study Reference Diagram Q9.

Explain why certain areas such as the USA, Europe and Japan use such a great amount of the world's oil.

Marks

4

DO NOT
WRITE IN
THIS
MARGIN

KU ES

[Turn over

10. **Reference Diagram Q10: Effects of Asian Tsunami, December 2004**

Damage in Nam Khem village, Thailand

Devastation at Petang beach resort, Thailand

Farmland destroyed, Sri Lanka

Marks

10. (continued)

Study Reference Diagram Q10.

> "Immediate help is essential but we also need long term aid for a full recovery."

Government spokesperson

For this type of natural disaster, **describe** what could be done to help these areas **in the longer term**.

_____

_____

_____

_____

_____

_____

4

[END OF QUESTION PAPER]

[BLANK PAGE]

STANDARD GRADE | CREDIT

2007

[BLANK PAGE]

C

1260/405

| NATIONAL QUALIFICATIONS 2007 | TUESDAY, 8 MAY 1.00 PM – 3.00 PM | GEOGRAPHY STANDARD GRADE Credit Level |

All questions should be attempted.

Candidates should read the questions carefully. Answers should be clearly expressed and relevant.

Credit will always be given for appropriate sketch-maps and diagrams.

Write legibly and neatly, and leave a space of about one centimetre between the lines.

All maps and diagrams in this paper have been printed in black only: no other colours have been used.

SCOTTISH QUALIFICATIONS AUTHORITY

©

1:50 000 Scale
Landranger Series

Four colours should appear ab
Four colours should appear ab

2 centimetre

1 kilometre = 0·6214 mile

Extract No 1558/54

DUNDEE

TAY

BROUGHTY
FERRY

Barnhill

TAYPORT

1 mile = 1·6093 kilometres

Magnetic North   Grid North   True North

Diagrammatic only

000
(one grid square)

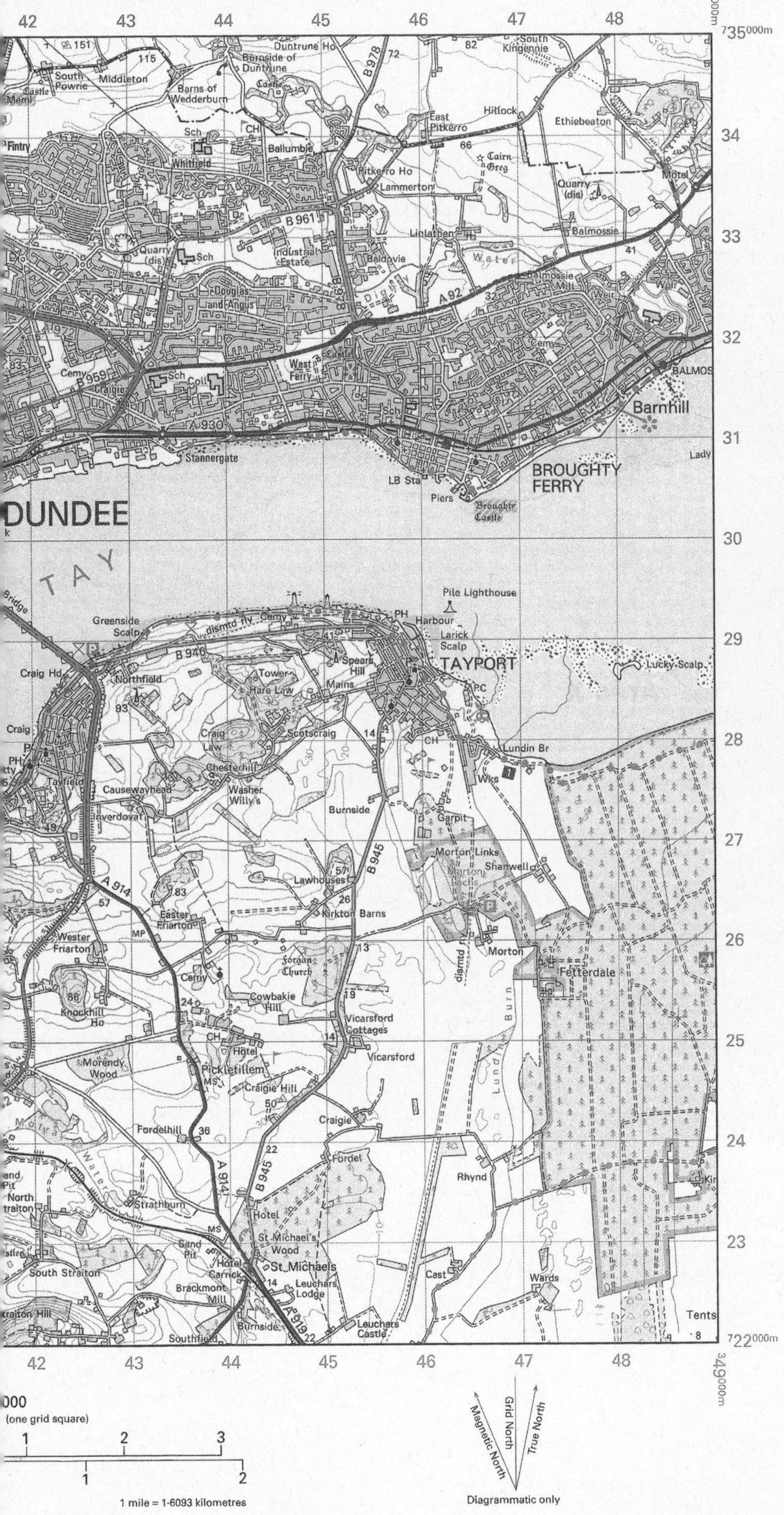

1.   **Reference Diagram Q1A**

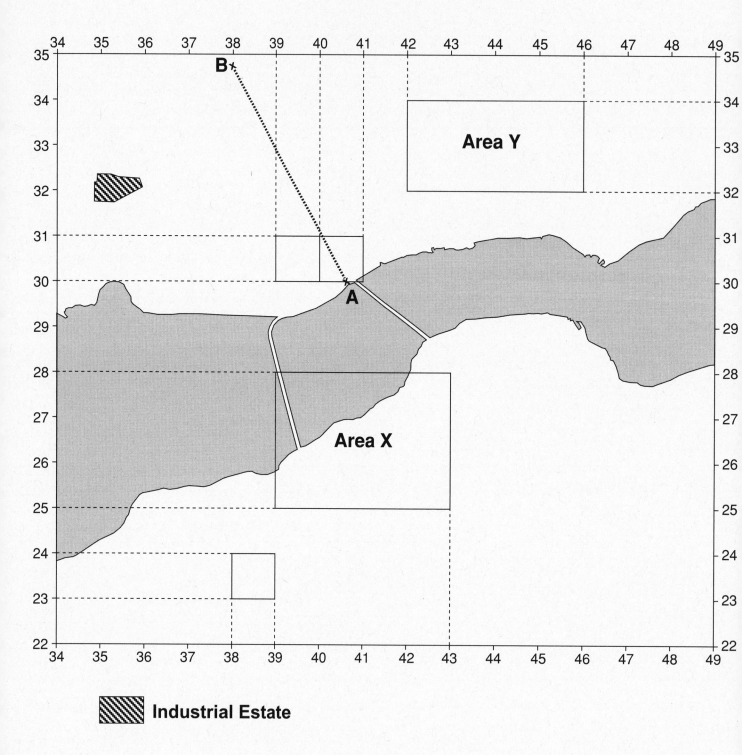

 Industrial Estate

Marks
KU | ES

1. (continued)

This question refers to the OS Map Extract (No 1558/54) of the Dundee area.

Reference Diagram Q1B: View SE from Dundee Law 392313

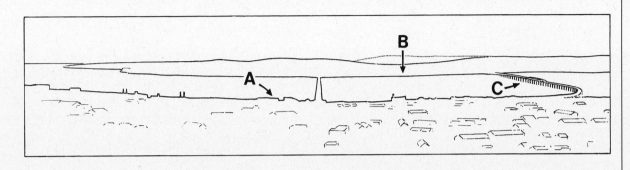

(a) Study Reference Diagram Q1B and the map extract.

Identify the **three** features A, B and C.

Choose from:

Railway Bridge;   Tayport;   Discovery Point;   Road Bridge;

Docks;   Newport on Tay.

3

(b) Find Area X on Reference Diagram Q1A (see *Page two*) and on the OS map extract.

Referring to map evidence, **explain** the way in which the **physical** landscape has affected land use in this area.

4

[Turn over

Mark

KU

1. (continued)

Reference Diagram Q1C: Area Y in the Year 1971

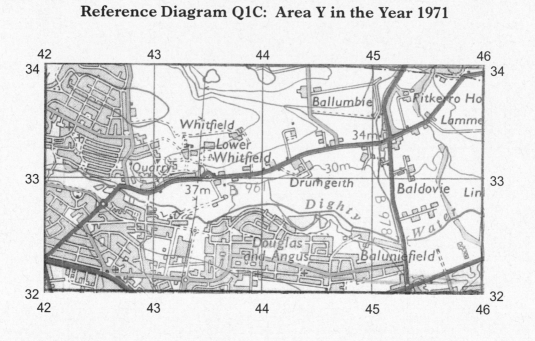

(*c*) Look at Reference Diagram Q1C.

Find Area Y on Reference Diagram Q1A and on the OS map extract.

Describe how the land use changes in the area since 1971 have **both** benefited **and** created problems for the area and its people.

(*d*) Mr Dick works in grid square 4030 and lives in a flat in grid square 3930. He is considering moving house to Gauldry 3823.

Do you think he should make this move?

Using map evidence, give reasons for your answer.

(*e*) Refer to Reference Diagram Q1A.

A group of students intends to gather information about differences in urban land use along transect AB (407300 to 380348).

Describe, in detail, the gathering techniques they might use.

Give reasons for your choice of techniques.

(*f*) **Explain** the location of the industrial estate at 3532.

4

Marks

KU | ES

2.   **Reference Diagram Q2: A Hanging Valley**

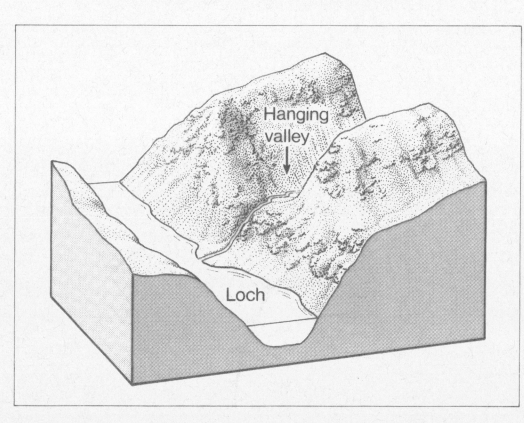

Look at Reference Diagram Q2.

Explain how a hanging valley is formed.

You may use diagram(s) to illustrate your answer.

4

[Turn over

Mark

KU

3. **Reference Diagram Q3A: Synoptic Chart 12 noon, 18 November 2006**

Reference Diagram Q3B: Two Sets of Weather Information

	Set X	**Set Y**
Temperature	5 °C	12 °C
Wind speed	35 knots	10 knots
Wind direction	SW	E
Precipitation	Heavy rain	Steady rain
Cloud amount	7 oktas	4 oktas
Cloud type	Cumulonimbus	Stratus

Look at Reference Diagrams Q3A and Q3B.

Which set of weather information, X or Y, is correct for Bristol?

Explain your choice in detail.

5

Marks

KU   ES

4.   **Reference Diagram Q4: Desertification**

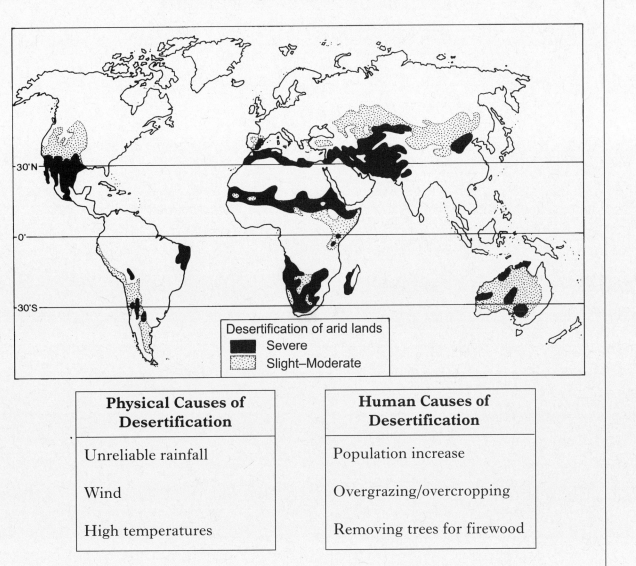

Physical Causes of Desertification	**Human Causes of Desertification**
Unreliable rainfall	Population increase
Wind	Overgrazing/overcropping
High temperatures	Removing trees for firewood

Look at Reference Diagram Q4.

(a) Desertification is a major problem in many areas of the world.

Choose **one** physical and **one** human cause and **explain** why each of them is a major reason for desertification.

4

(b) **Describe**, in detail, ways in which desertification can be overcome.

4

[Turn over

Marks

KU

5. **Reference Diagram Q5: Central Business District of a Large City**

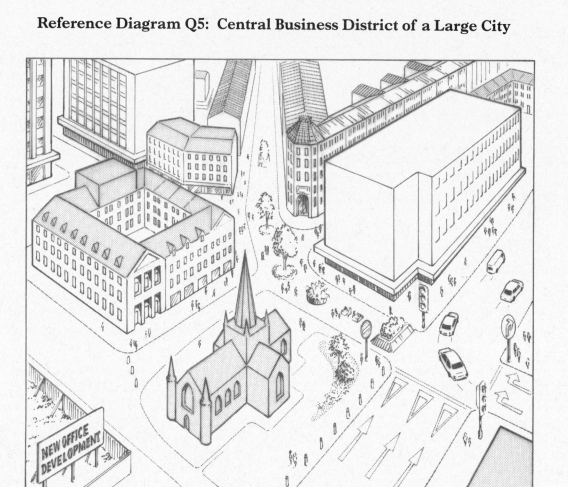

Study Reference Diagram Q5.

In recent years many changes have taken place in the Central Business Districts of British cities.

Give reasons for these changes.

5

6. **Reference Diagram Q6A: Inverlochlarig Sheep Farm, Perthshire**

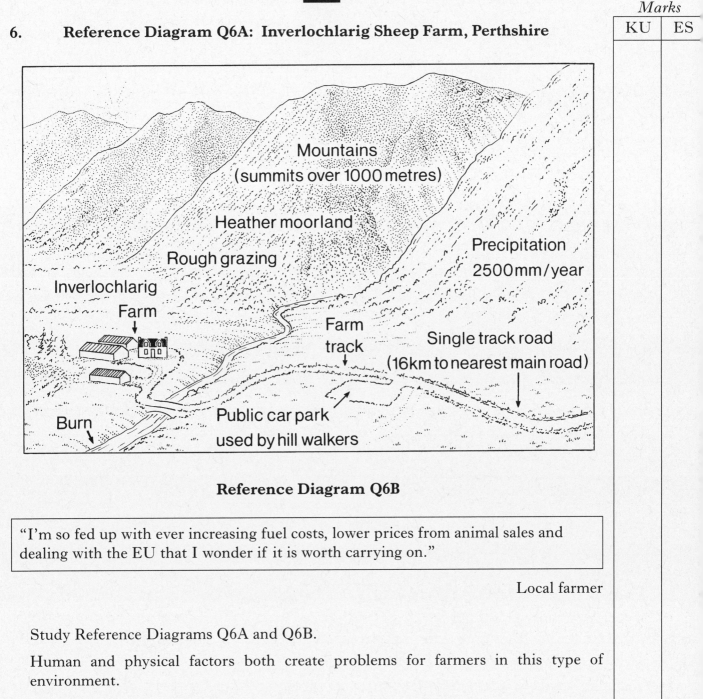

Mountains
(summits over 1000 metres)

Heather moorland

Rough grazing

Inverlochlarig
Farm

Farm
track

Precipitation
2500 mm/year

Single track road
(16 km to nearest main road)

Public car park
used by hill walkers

Burn

Reference Diagram Q6B

"I'm so fed up with ever increasing fuel costs, lower prices from animal sales and dealing with the EU that I wonder if it is worth carrying on."

Local farmer

Study Reference Diagrams Q6A and Q6B.

Human and physical factors both create problems for farmers in this type of environment.

Select **either** human **or** physical factors.

For the factors you have chosen **explain**, in detail, why these cause more problems for the farmer.

6

[Turn over

Reference Diagram Q7A:
The Eden Project—A Visitor Attraction in Cornwall

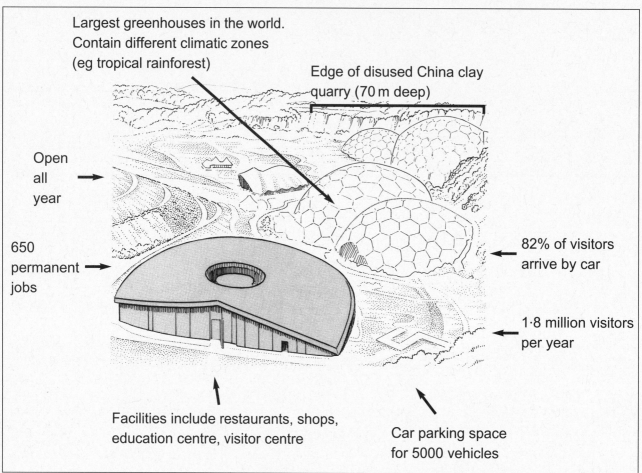

Largest greenhouses in the world.
Contain different climatic zones
(eg tropical rainforest)

Edge of disused China clay
quarry (70 m deep)

Open
all
year

82% of visitors
arrive by car

650
permanent
jobs

1·8 million visitors
per year

Facilities include restaurants, shops,
education centre, visitor centre

Car parking space
for 5000 vehicles

Reference Diagram Q7B: Location of the Eden Project

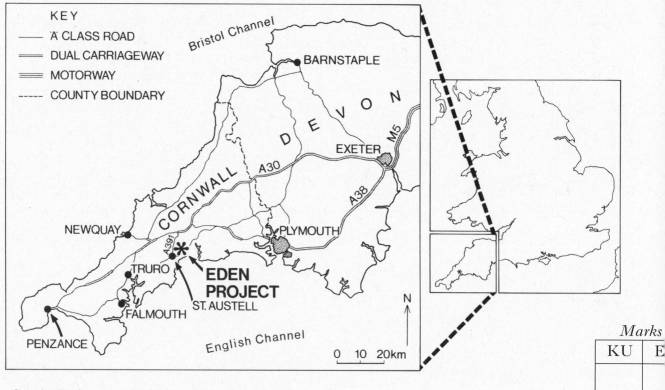

KEY
— 'A' CLASS ROAD
= DUAL CARRIAGEWAY
≡ MOTORWAY
---- COUNTY BOUNDARY

Bristol Channel

BARNSTAPLE

D E V O N

EXETER   M5

CORNWALL   A30

A38

NEWQUAY   A39

PLYMOUTH

TRURO   ✱

EDEN
PROJECT
ST. AUSTELL

FALMOUTH

PENZANCE

English Channel

N

0   10   20km

Marks

KU   E

Study Reference Diagrams Q7A and Q7B.

Explain fully the advantages **and** disadvantages of this new visitor attraction to St Austell and the surrounding area.

8. **Reference Diagram Q8A:**
   North America Population Distribution

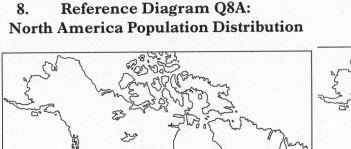

Reference Diagram Q8B:
North America Relief

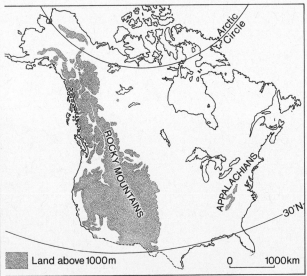

Reference Diagram Q8C:
North America Annual Rainfall

Reference Diagram Q8D:
North America Power and Industry

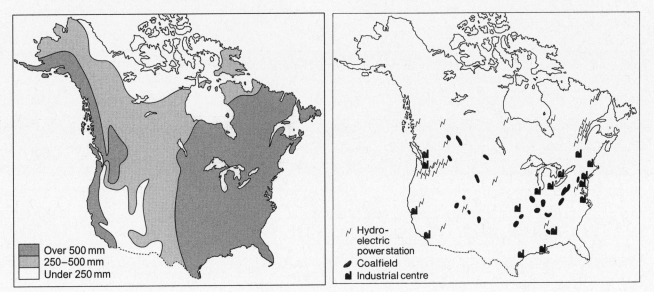

	Marks	
	KU	ES

Using information given in Reference Diagrams Q8A, B, C and D, **explain** the distribution of population in North America.

6

[Turn over

9.   **Reference Diagram Q9A: Tourism in the Gambia (West Africa)**

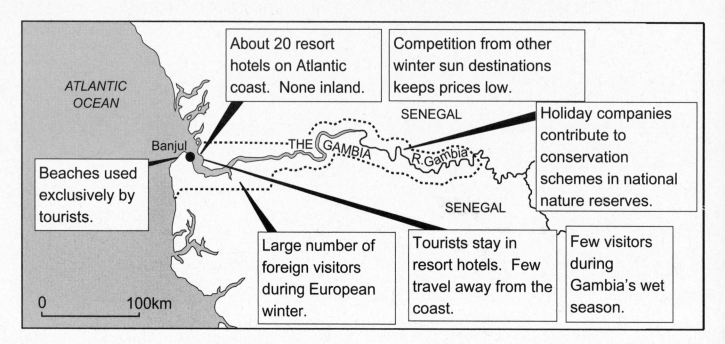

Marks

KU	E

(*a*) The Gambia which is a country in West Africa is known as a "winter sun" destination for tourists from Europe.

"The growth of the tourist industry has brought huge benefits for the people and the environment of the Gambia."

Package holiday company spokesperson

Do you agree with this statement?

Give reasons for your answer.

Marks

KU	ES

9. (continued)

Reference Diagram Q9B: Tourism Facts for The Gambia

- Population   =   1 400 000

- Labour Force =   400 000

- Tourism employs 10 000 Gambians

- Tourism is the biggest foreign exchange earner in The Gambia

- GDP   from Agriculture = 30%   from Tourism = 25%

      from Fisheries   = 30%   from Others   = 15%

- Most visitors come from Europe:

      from UK   = 60%   from Sweden = 7%

      from Netherlands = 12%   from Others   = 21%

(*b*) Study Reference Diagram Q9B.

Which processing techniques would be most effective to show the different information about tourism?

Explain your choice of techniques.

4

[Turn over for Question 10 on *Page fourteen*

Mark
KU

10.   **Reference Diagram Q10: Measures of Development**

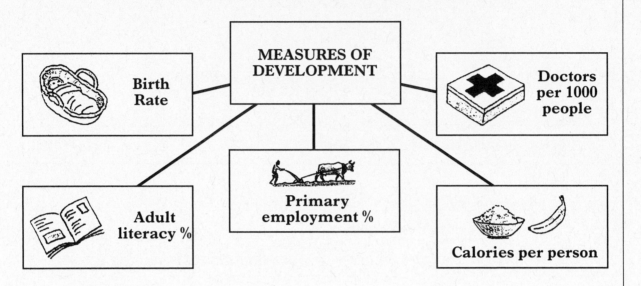

Study Reference Diagram Q10.

Choose **two** of the measures shown and **explain** why they are good indicators of the differences between ELDCs (economically less developed countries) and EMDCs (economically more developed countries).

6

[END OF QUESTION PAPER]

STANDARD GRADE | GENERAL

2008

[BLANK PAGE]

G

FOR OFFICIAL USE

KU	ES

Total Marks

1260/403

NATIONAL
QUALIFICATIONS
2008

FRIDAY, 9 MAY
10.25 AM–11.50 AM

GEOGRAPHY
STANDARD GRADE
General Level

Fill in these boxes and read what is printed below.

Full name of centre

Town

Forename(s)

Surname

Date of birth

| Day | Month | Year | | Scottish candidate number | | Number of seat |

1 Read the whole of each question carefully before you answer it.

2 Write in the spaces provided.

3 Where boxes like this ☐ are provided, put a tick ✓ in the box beside the answer you think is correct.

4 Try all the questions.

5 Do not give up the first time you get stuck: you may be able to answer later questions.

6 Extra paper may be obtained from the invigilator, if required.

7 Before leaving the examination room you must give this book to the invigilator. If you do not, you may lose all the marks for this paper.

Extract No 1655/36

1:50 000 Scale
Landranger Series

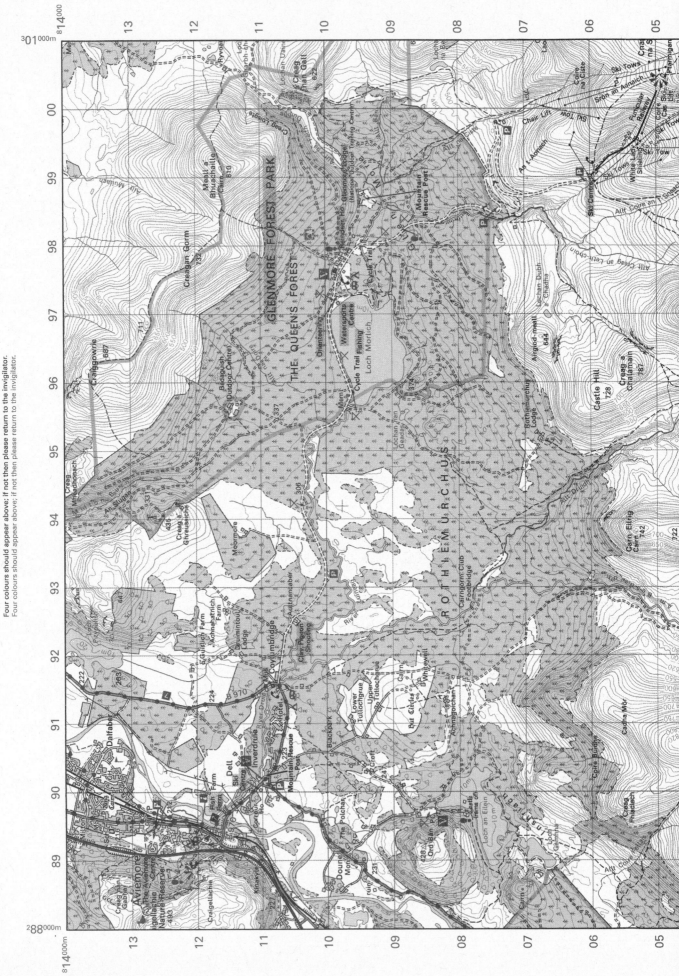

Scale 1:50 000

2 centimetres to 1 kilometre (one grid square)

1 kilometre = 0·6214 mile

1 mile = 1·6093 kilometres

1.   **Reference Diagram Q1A: The Aviemore Area**

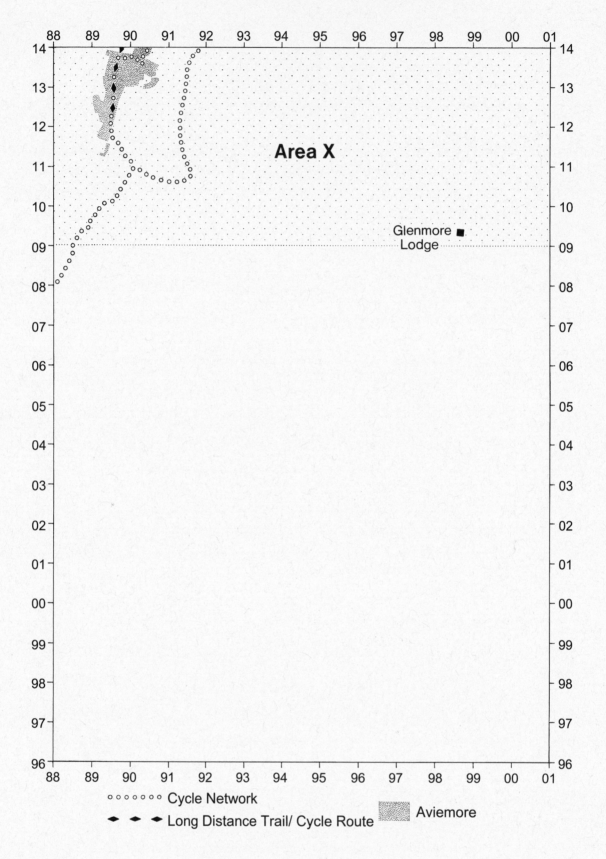

Marks

1. (continued)

Look at the Ordnance Survey Map Extract (No 1655/36) of the Aviemore area and Reference Diagram Q1A on Page two.

(*a*) Using the map extract, match the glaciated features A, B, C and D to the correct name/grid reference in the table below.

A Ribbon Lake   B Corrie   C U shaped Valley   D Pyramidal Peak

Grid Reference/Name of Feature	Letter
9597 Angel's Peak	
9798 Lairig Ghru	
9400 Loch Coire an Lochain	
9198 Loch Einich	

3

(*b*) **Explain** how a pyramidal peak was formed.

You may use diagram(s) to illustrate your answer.

_____

_____

_____

_____

_____

_____

_____

_____

_____

_____

3

Marks

1. (continued)

(*c*)   Find Area X on the map extract and on Reference Diagram Q1A.

In what ways has the physical landscape in Area X both encouraged **and** limited settlement growth?

_____

_____

_____

_____

_____

4

(*d*)   Glenmore Lodge (9809) is a National Outdoor Training Centre*.

(*This type of centre provides instruction in outdoor activities.)

How good is this location for the centre?

Give reasons for your answer.

_____

_____

_____

_____

_____

4

Marks

1. (continued)

Reference Diagram Q1B: Selected Aims of National Parks

> • *Preserve beauty of countryside*
>
> • *Conserve local wildlife*
>
> • *Provide good access and facilities for public enjoyment*
>
> • *Maintain farming*

(*e*) This area is part of the Cairngorm National Park.

Explain how land uses shown on the OS map **and** outdoor activities in this area might be in conflict with the aims shown in Reference Diagram Q1B.

_____

_____

_____

_____

_____

4

(*f*) There is a Long Distance Trail and Cycle Network shown on both the map extract and Reference Diagram Q1A.

What techniques could be used to gather information on the impact of walkers and cyclists on the local area?

Give reasons for your choices.

_____

_____

_____

_____

_____

4

[Turn over

Marks

2.   Reference Diagram Q2: A Lowland River Landscape

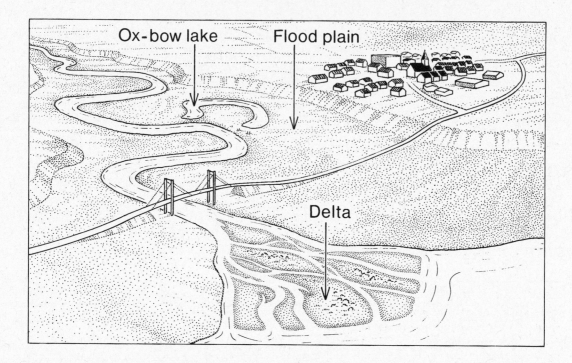

Look at Reference Diagram Q2.

Choose **one** of the named river features shown and **explain** how it was formed.

You may use a diagram(s) to illustrate your answer.

_____

_____

_____

_____

_____

_____

_____

_____

3

Marks

3. **Reference Diagram Q3A:**
Weather Station Symbol for
Aberdeen 12 noon, 17th December

Reference Diagram Q3B:
Advertisement for
Football Match

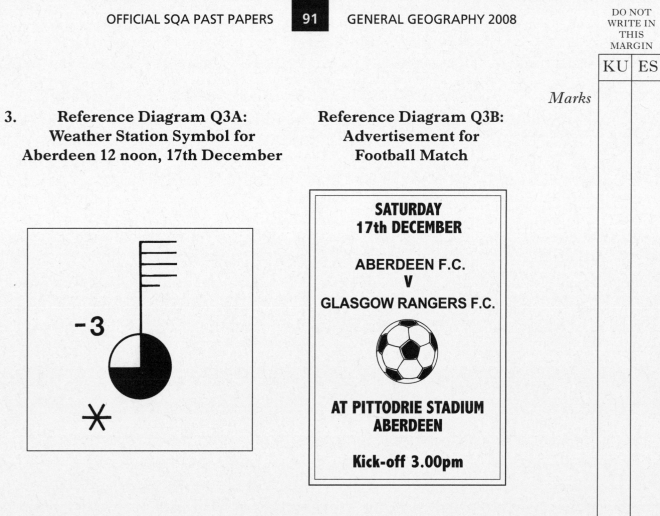

Look at Reference Diagrams Q3A and Q3B.

On Saturday morning the referee decided to postpone this game.

Referring to the weather conditions, give reasons for his decision.

_____

_____

_____

_____

_____

4

[Turn over

Marks

4.   **Reference Diagram Q4A: Causes of Sea Pollution**

A	Sewage and industrial waste	40%
B	Wind-blown gases and particles from industry	35%
C	Oil spills from tankers	15%
D	Dumping at sea	10%

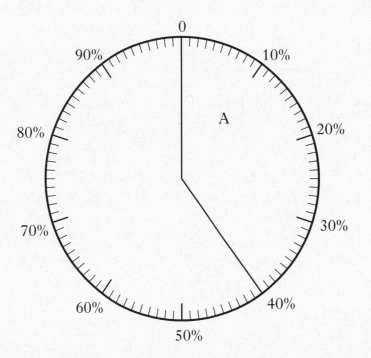

(*a*)   Complete the pie chart to show the information given in Reference Diagram Q4A.

3

DO NOT
WRITE IN
THIS
MARGIN

KU | ES

4. (continued)

Reference Diagram Q4B: Methods of reducing Sea Pollution

A	Enforce laws banning dumping at sea
B	Ensure that sewage is treated before it goes out to sea

Marks

(b)   Which **one** of the above measures do you think would be the best method of reducing sea pollution?

Choice: **A** or **B** _____

Give reasons for your choice.

_____

_____

_____

_____

_____

_____

3

[Turn over

5.   **Reference Diagram Q5A: Selected Climate Regions**

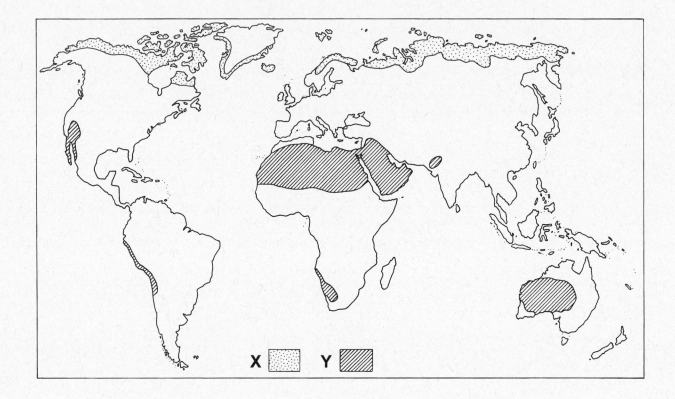

5. (continued)

Reference Diagram Q5B: Climate Graphs for Selected Regions

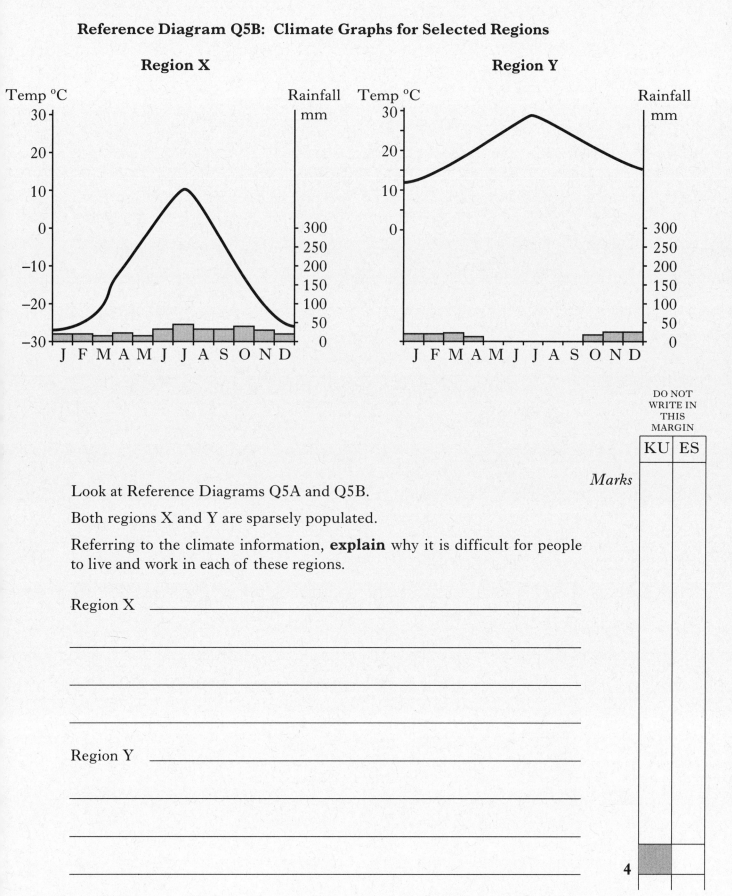

Look at Reference Diagrams Q5A and Q5B.

Both regions X and Y are sparsely populated.

Referring to the climate information, **explain** why it is difficult for people to live and work in each of these regions.

Region X _____

_____

_____

_____

Region Y _____

_____

_____

_____

KU	ES

Marks

4

Marks

6.   **Reference Diagram Q6: Selected Farm Data**

Average Temperature	January 5°C, July 16°C
Annual Precipitation	600–800 mm
Relief	Flat, gently sloping land
Soils	Alluvial soils on floodplain
Location	5 km to nearest town

Look at Reference Diagram Q6.

The conditions shown might be suitable for either dairy farming or arable farming.

Which type of farming do you think is more likely?

Tick (✓) your choice.

Dairy farming ☐   Arable farming ☐

Give reasons for your choice.

_____

_____

_____

_____   4

KU | ES

Marks

7. **Reference Diagram Q7: A Modern Industrial Landscape**

Near edge of town or city

No chimneys

Large areas of tarmac surface

Landscaped with grass, trees and shrubs

Close to main roads and motorways

Spacious site on flat land

Look at Reference Diagram Q7.

Explain why some of the labelled features are typical of a modern industrial landscape.

_____

_____

_____

_____

_____

_____

4

[Turn over

Marks

8.   **Reference Diagram Q8: Reducing Traffic Congestion**

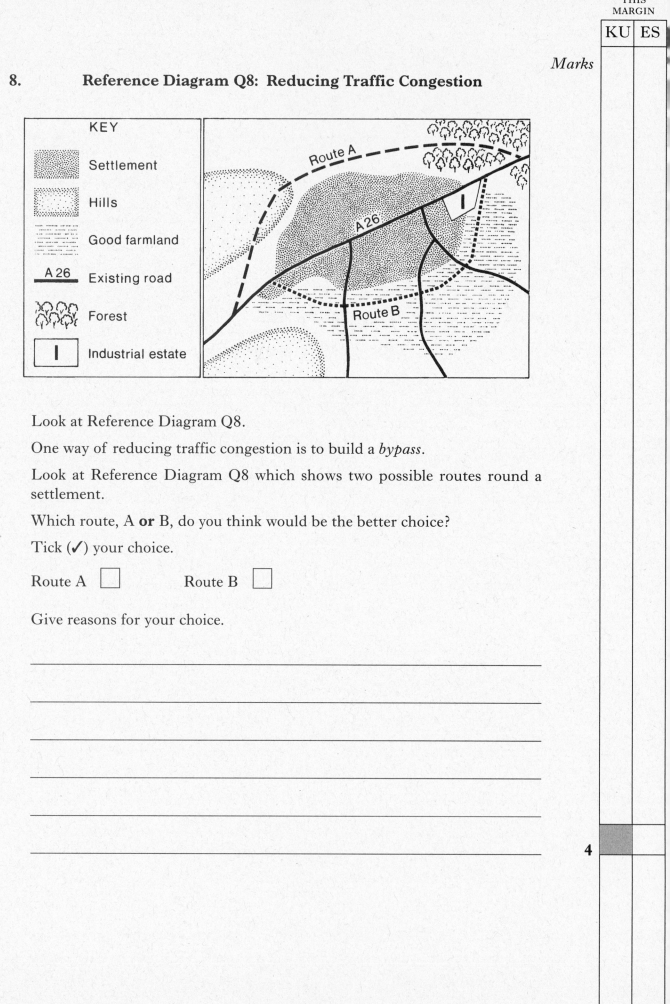

Look at Reference Diagram Q8.

One way of reducing traffic congestion is to build a *bypass*.

Look at Reference Diagram Q8 which shows two possible routes round a settlement.

Which route, A **or** B, do you think would be the better choice?

Tick (✓) your choice.

Route A ☐   Route B ☐

Give reasons for your choice.

_____

_____

_____

_____

_____

_____

4

[Turn over for Question 9 on *Page sixteen*

Marks

9.   **Reference Diagram Q9A: Factors which influence
Death Rates in Europe**

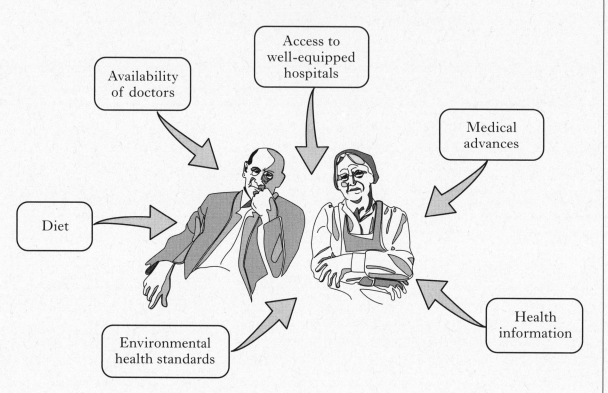

(*a*)   Study Reference Diagram Q9A.

For any **two** of the factors shown, **explain** how they affect death rates in Europe.

Factor 1 _____

_____

_____

_____

_____

Factor 2 _____

_____

_____

_____

_____

4

KU | ES

Marks

9. (continued)

Reference Diagram Q9B: Life Expectancy in Italy

Year	Life Expectancy
1950	64
1960	66
1970	70
1980	72
1990	75
2000	78
2005	79

(*b*)   Look at Reference Diagram Q9B.

Name **two** other techniques which could be used to show the information given.

Give reasons for your choices.

Technique 1 _____

Reason(s) _____

_____

_____

_____

Technique 2 _____

Reason(s) _____

_____

_____

_____

4

[Turn over

Marks

10.   **Reference Diagram Q10A: Tied Aid**

Allows farming
to be improved

Money to be used to buy
goods from EMDC**

To be used for
projects identified
by EMDC**

TIED AID

Given by EMDC
to ELDC*

Loans to be
paid back with interest

*ELDC = Economically Less Developed Country
**EMDC = Economically More Developed Country

(*a*)   Look at Reference Diagram Q10A.

What are the advantages **and** disadvantages of tied aid for ELDCs?

_____

_____

_____

_____

_____   4

Marks

10. (continued)

Reference Diagram Q10B: Education in ELDCs

(*b*) Look at Reference Diagram Q10B.

"Education is the best way to improve living conditions in ELDCs."

Do you agree fully with this statement?

Give reasons for your answer.

_____

_____

_____

_____

_____

_____   **3**

[Turn over for Question 11 on *Page twenty*

KU | ES

Marks

11. **Reference Diagram Q11: Location of UK and India**

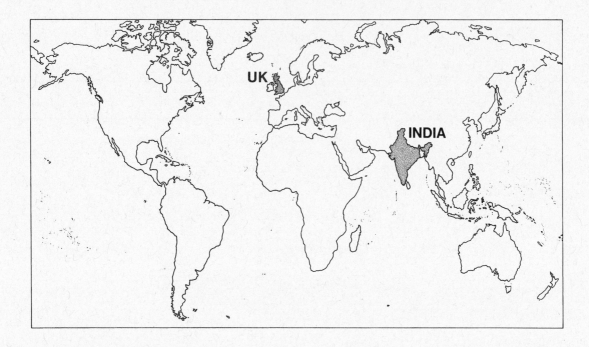

The following statements are about Development and Trade in the UK or India.

Match the letters to the correct country in the table below.

A   Quotas on exports to Germany and France

B   Exports mainly manufactured goods

C   Agriculture employs 66% of population

D   Energy consumption per capita is low

E   Agriculture employs 2% of population

F   High GNP per capita

UK	India

4

[END OF QUESTION PAPER]

[BLANK PAGE]

C

1260/405

NATIONAL QUALIFICATIONS 2008	FRIDAY, 9 MAY 1.00 PM – 3.00 PM	**GEOGRAPHY** STANDARD GRADE Credit Level

All questions should be attempted.

Candidates should read the questions carefully. Answers should be clearly expressed and relevant.

Credit will always be given for appropriate sketch-maps and diagrams.

Write legibly and neatly, and leave a space of about one centimetre between the lines.

All maps and diagrams in this paper have been printed in black only: no other colours have been used.

Extract No 1656/104

1:50 000 Scale
Landranger Series

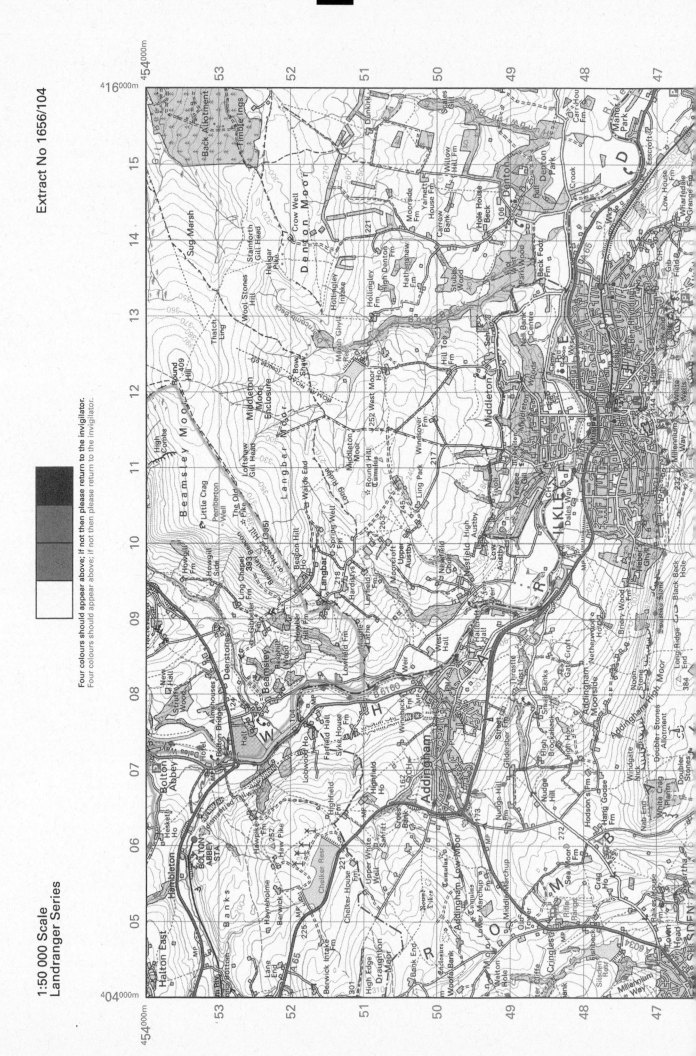

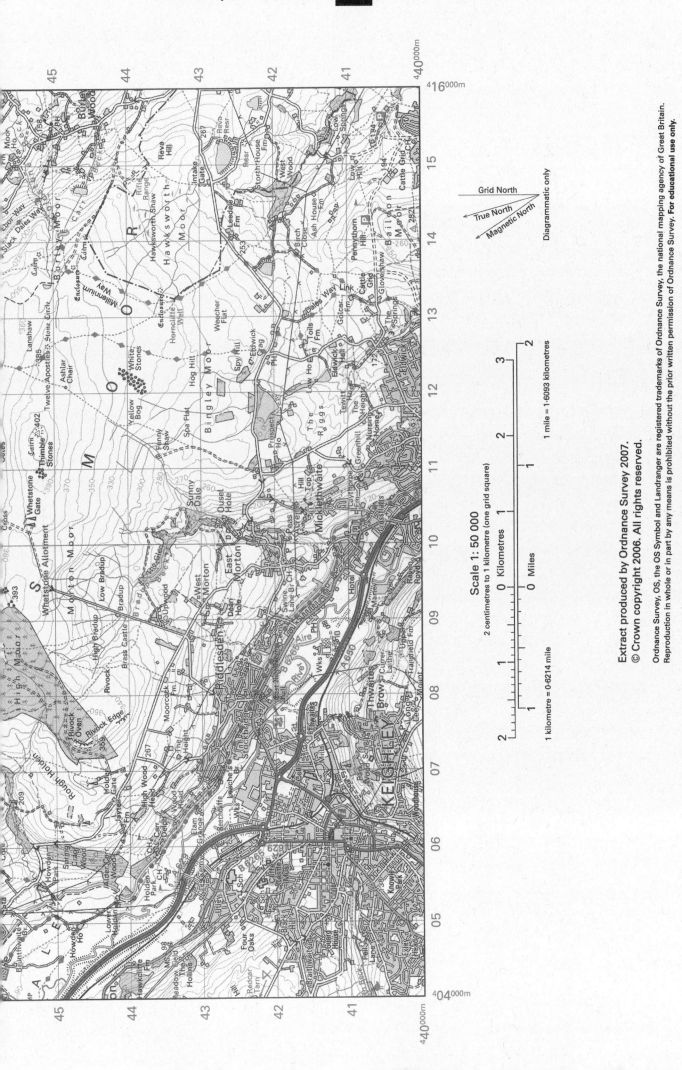

Scale 1: 50 000

2 centimetres to 1 kilometre (one grid square)

1 kilometre = 0·6214 mile

1 mile = 1·6093 kilometres

Extract produced by Ordnance Survey 2007.
© Crown copyright 2006. All rights reserved.

Grid North
True North
Magnetic North

Diagrammatic only

1.

Reference Diagram Q1

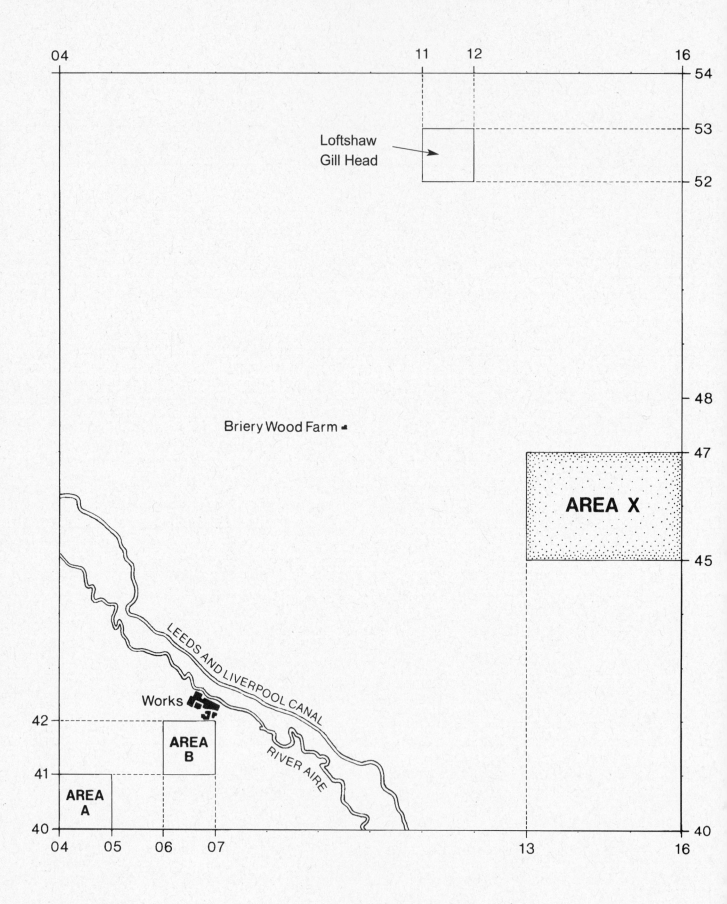

Marks

KU	ES

1. (continued)

This question refers to the OS Map Extract (No 1656/104) of the Ilkley area and Reference Diagram Q1.

(a) Describe the **physical** features of the River Aire **and** its valley from 041450 to 099400.

Your answer should **not** refer to the Leeds and Liverpool canal.

4

(b) The valley which runs south east from Loftshaw Gill Head in square 1152 is a "V" shaped river valley. Explain how this is likely to have been formed.

You may use diagrams to illustrate your answer.

4

(c) Find Briery Wood Farm at 095474.

What are the advantages **and/or** disadvantages of this location for a farm?

4

(d) Why might there be conflicts between the various land uses in Area X?

6

(e) Describe the differences between the urban environments of Area A (0440) and Area B (0641).

5

(f) Find the works in grid square 0642.

Why is this a good site for the works?

5

[Turn over

2.   **Reference Diagram Q2: A Lowland Landscape in Scotland**

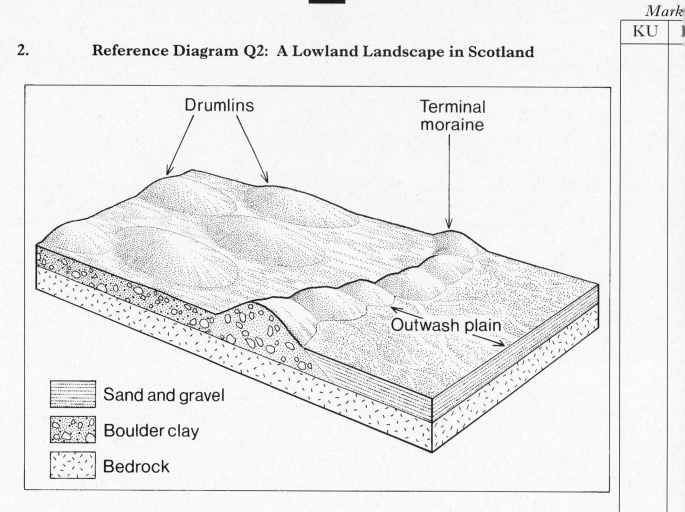

Look at Reference Diagram Q2.

Choose any **two** of the features shown in the diagram and **explain** how they were formed.

You may use diagram(s) to illustrate your answer.

5

3. **Reference Diagram Q3: European Synoptic Chart for noon, 8th July**

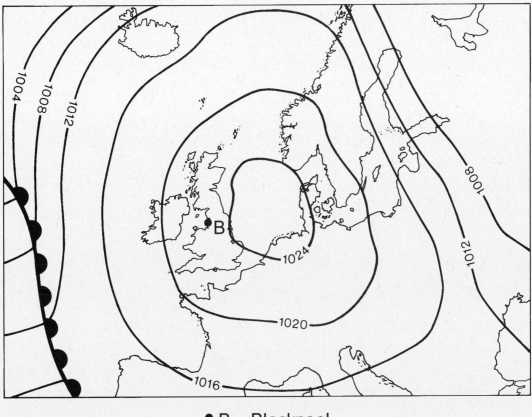

• B – Blackpool

Look at Reference Diagram Q3.

On July 8th Mr McCormack is taking his young family to Blackpool for a one week holiday.

Do you think the weather conditions will be favourable for them?

Give reasons your answer.

[Turn over   5

4.　　　**Reference Diagram Q4:　Brazilian Rainforest Facts**

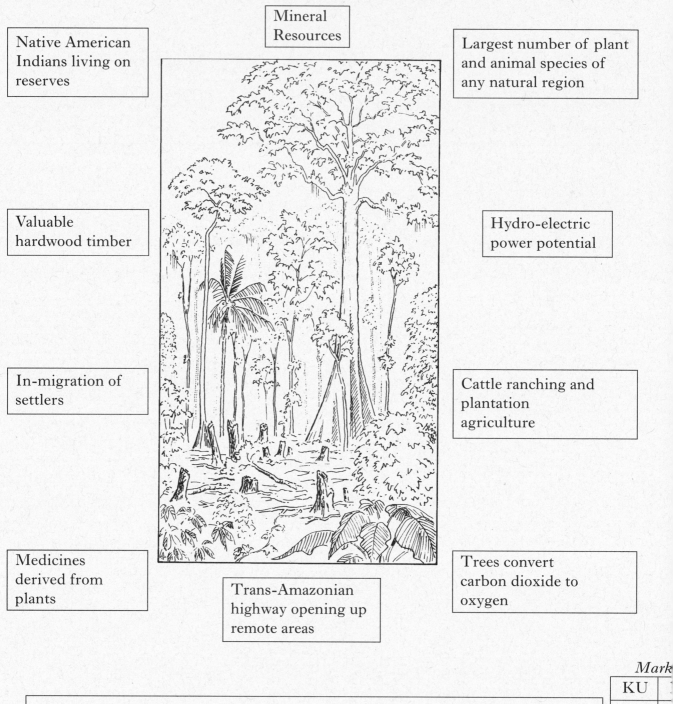

Mineral Resources

Native American Indians living on reserves

Largest number of plant and animal species of any natural region

Valuable hardwood timber

Hydro-electric power potential

In-migration of settlers

Cattle ranching and plantation agriculture

Medicines derived from plants

Trees convert carbon dioxide to oxygen

Trans-Amazonian highway opening up remote areas

Mark

KU

"Cutting down the rainforest in Brazil will affect the whole world more than it will affect Brazil itself."

(Statement by an environment spokesperson)

Look at Reference Diagram Q4 and the statement above.

To what extent do you agree with the statement?

Give reasons for your answer.

5.   **Reference Diagram Q5A: Location of Housing Types**

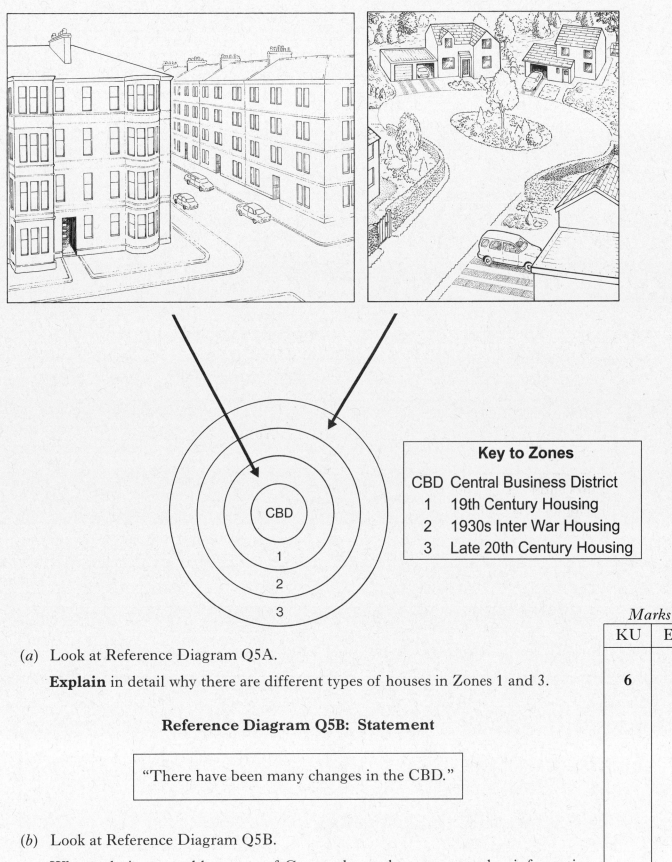

Key to Zones

CBD Central Business District
1   19th Century Housing
2   1930s Inter War Housing
3   Late 20th Century Housing

	Marks
	KU	ES

(a) Look at Reference Diagram Q5A.

Explain in detail why there are different types of houses in Zones 1 and 3.   **6**

Reference Diagram Q5B: Statement

"There have been many changes in the CBD."

(b) Look at Reference Diagram Q5B.

What techniques could a group of Geography students use to gather information on changes in the Central Business District (CBD)?

Give reasons for your chosen techniques.   **5**

6.   **Reference Diagram Q6A: Eurocentral**

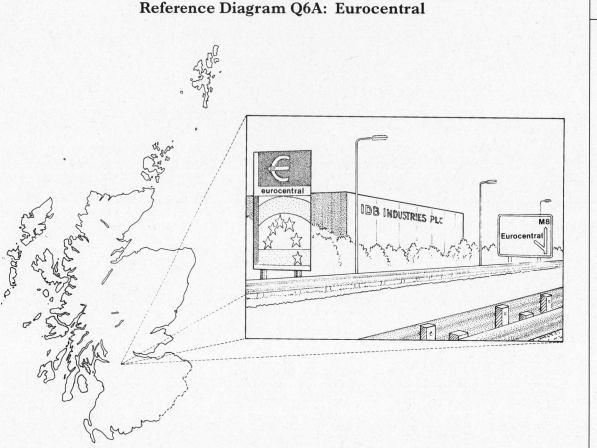

Reference Diagram Q6B: Extract from a News Report

"Eurocentral is a 260 hectare greenfield site where new large-scale industrial development is planned.

This is expected to have a significant impact on the surrounding communities, where old industry has been in decline."

(*a*) Look at Reference Diagrams Q6A and Q6B.

Describe the benefits **and** problems which new, large-scale industrial development may bring to areas such as this.

6

6. (continued)

Reference Diagram Q6C: North Lanarkshire—Selected Industrial Statistics

Table 1: Employment Categories

Public and other services	30 700
Retailing and wholesale	24 800
Manufacturing	18 700
Finance and business	14 500

Table 2: Unemployment 1996–2002

1996	12 500
1997	10 500
1998	10 000
1999	8500
2000	8000
2001	7500
2002	7000

(b) Look at Reference Diagram Q6C.

What other techniques could be used to present the data shown above?

Give reasons for your choices.

5

[Turn over

7.   **Reference Diagram Q7A: Demographic Transition Model**

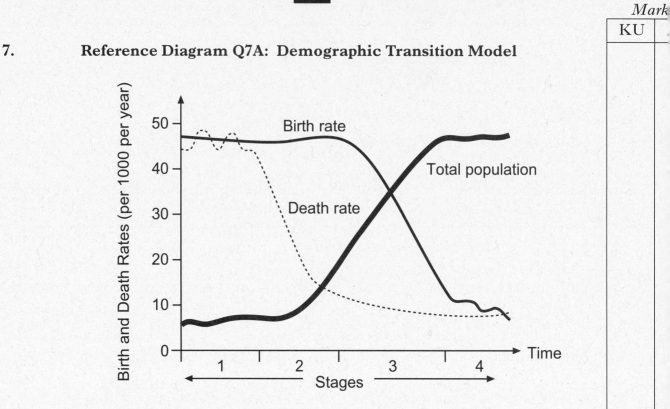

(*a*) Look at Reference Diagram Q7A.

Describe in detail the changes shown on the Demographic Transition Model from Stage 1 to Stage 4.

Reference Diagram Q7B: Selected Population Data

Country	Crude Birth Rate per 1000	Crude Death Rate per 1000	Natural Increase per 1000
India	23	8	15
Nigeria	38	14	24
UK	10·8	10·1	0·7
USA	14·1	8·3	5·8

(*b*) Look at Reference Diagram Q7B.

Choose **one** country shown on the table.

Suggest reasons for its **rate of natural increase**.

4

Marks

| KU | ES |

8.

Reference Diagram Q8A: Japan's Exports

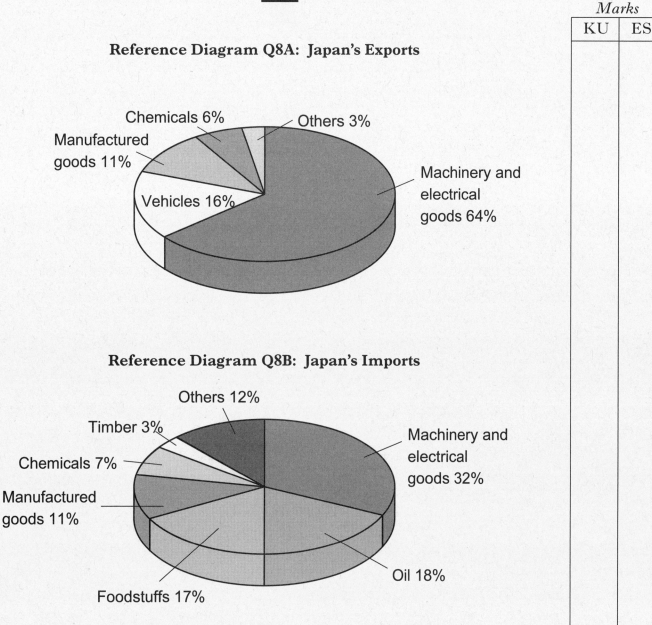

Reference Diagram Q8B: Japan's Imports

Look at Reference Diagrams Q8A and Q8B.

(*a*) What are the main differences between Japan's exports and imports?

3

(*b*) Japan has a large trade surplus and is the world's second biggest trading nation.

   Explain why Japan depends on world trade for the success of its economy.

3

[END OF QUESTION PAPER]

[BLANK PAGE]

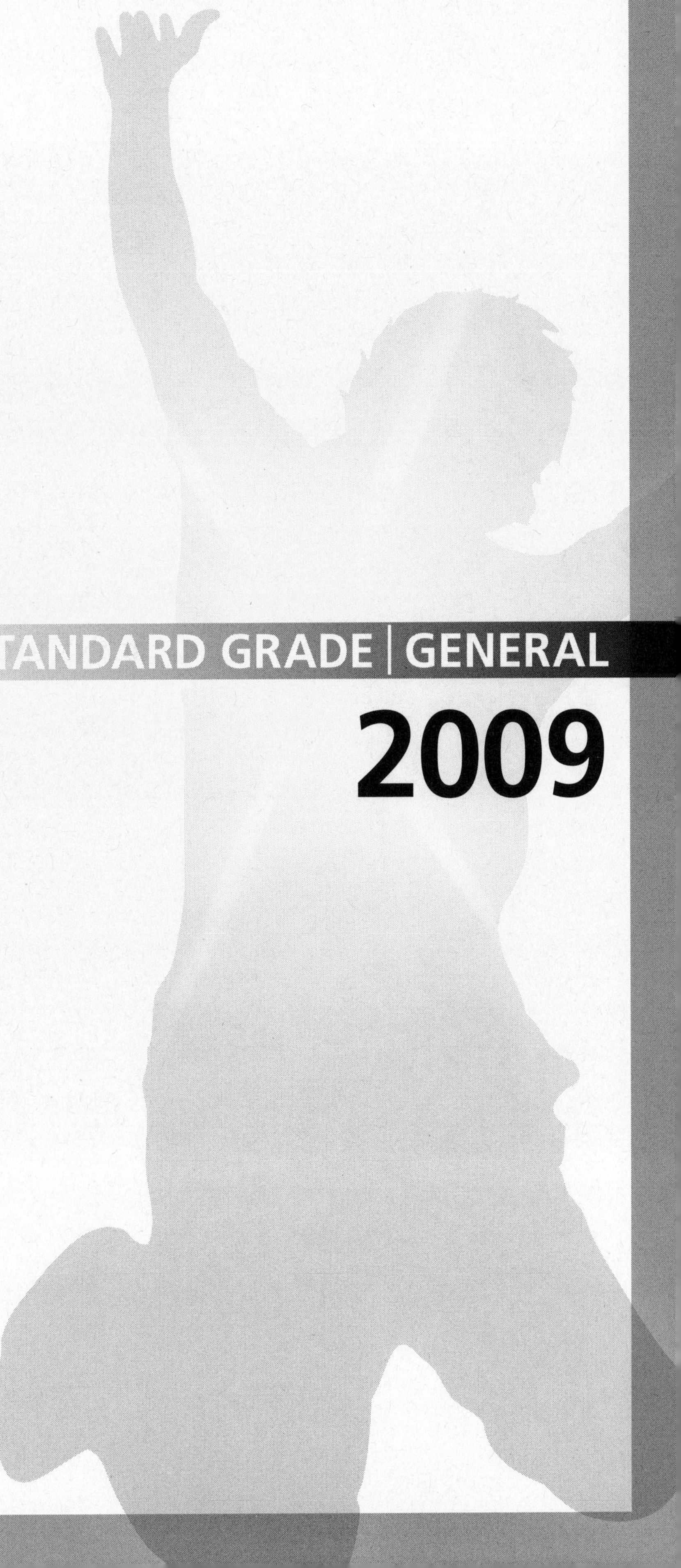

STANDARD GRADE | GENERAL

2009

[BLANK PAGE]

FOR OFFICIAL USE

KU	ES

Total Marks

1260/403

NATIONAL
QUALIFICATIONS
2009

WEDNESDAY, 27 MAY
10.25 AM–11.50 AM

GEOGRAPHY
STANDARD GRADE
General Level

Fill in these boxes and read what is printed below.

Full name of centre

Town

Forename(s)

Surname

Date of birth
Day Month Year   Scottish candidate number   Number of seat

1 Read the whole of each question carefully before you answer it.

2 Write in the spaces provided.

3 Where boxes like this ☐ are provided, put a tick ✓ in the box beside the answer you think is correct.

4 Try all the questions.

5 Do not give up the first time you get stuck: you may be able to answer later questions.

6 Extra paper may be obtained from the invigilator, if required.

7 Before leaving the examination room you must give this book to the invigilator. If you do not, you may lose all the marks for this paper.

Extract No 1742/140

1:50 000 Scale
Landranger Series

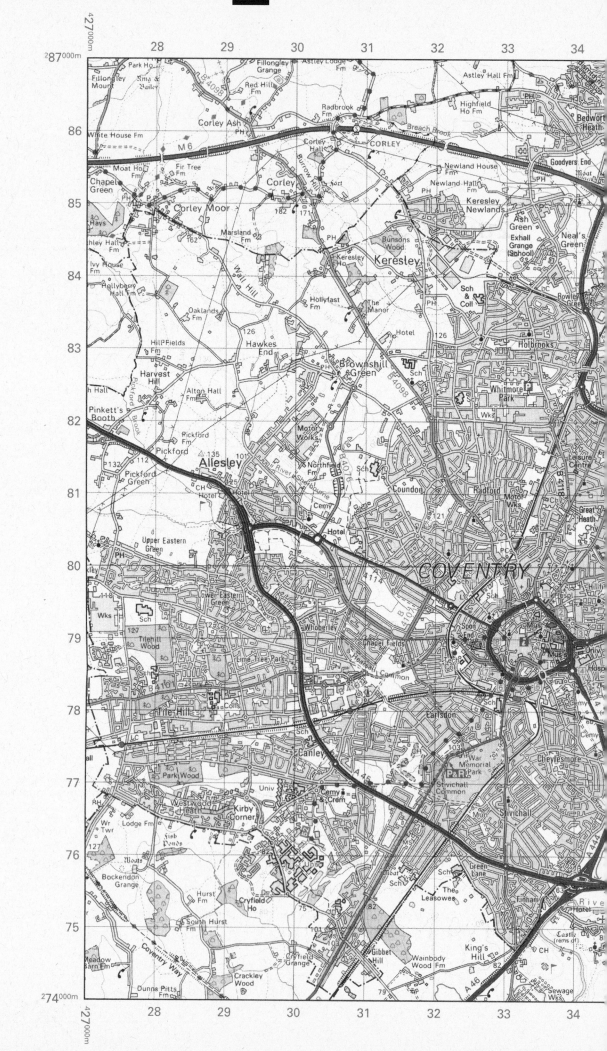

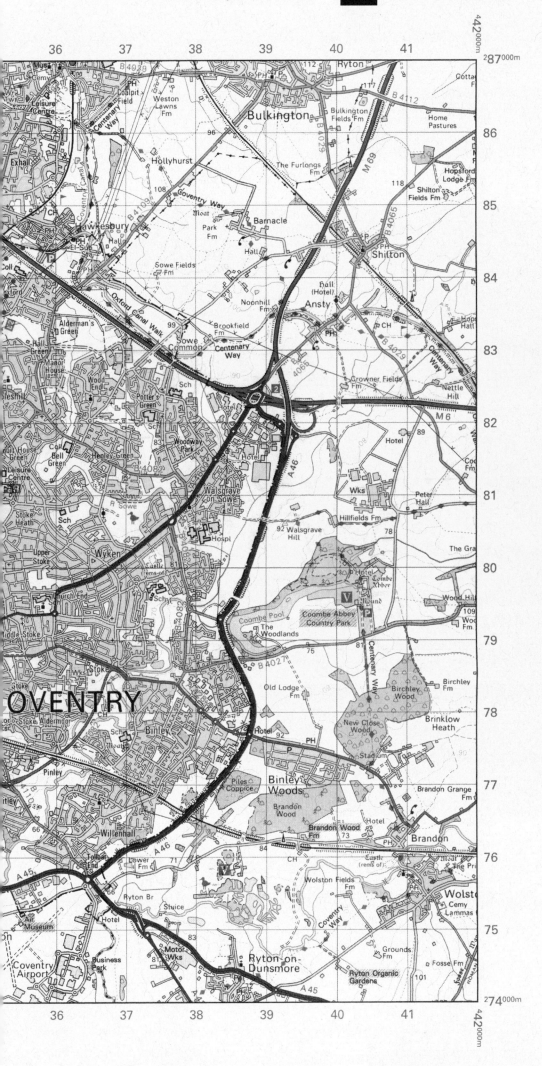

Scale 1: 50 000

2 centimetres to 1 kilometre (one grid square)

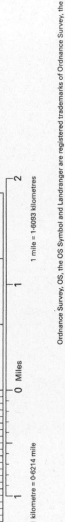

Grid North
Magnetic North
True North

Diagrammatic only

1.   Reference Diagram Q1A

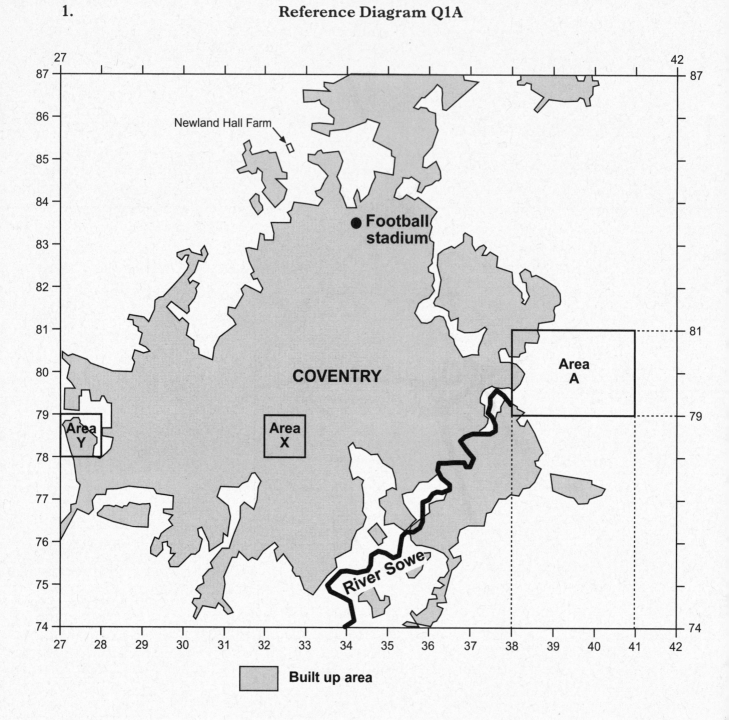

Built up area

Marks

1. (continued)

Look at the Ordnance Survey Map Extract (No 1742/140) of Coventry and Reference Diagram Q1A on *Page two*.

(a) Describe the **physical** features of the River Sowe **and** its valley from 378796 to 340740. (This section of the river is shown on Reference Diagram Q1A.)

_____

_____

_____

_____

_____

_____

4

(b) Find Area A on Reference Diagram Q1A and the map extract.

What are the advantages **and** disadvantages of this area for a Country Park?

_____

_____

_____

_____

_____

4

[Turn over

DO NOT
WRITE IN
THIS
MARGIN

Marks

1. (continued)

(*c*) Give map evidence to show that part of Coventry's Central Business District (CBD) is located in grid square 3379.

_____

_____

_____

_____

_____

_____

4

(*d*) A young couple want to buy a house in Coventry. They have decided to buy a house in either Area X (grid square 3278) or in Area Y (grid square 2778).

Using map evidence, which area, X or Y, would you advise them to choose?

Give reasons for your answer.

Choice: _____

Reasons _____

_____

_____

_____

_____

4

Marks

1. (continued)

(*e*)   Newland Hall Farm is located in map square 3285.

What are the advantages **and** disadvantages of this location for farming?

Advantages _____

_____

_____

_____

Disadvantages _____

_____

_____

_____

4

[Turn over

1. (continued)

Reference Diagram Q1B: Land Use at Grid Reference 343835 in 1999

Reference Diagram Q1C: Land Use at Grid Reference 343835 in 2009

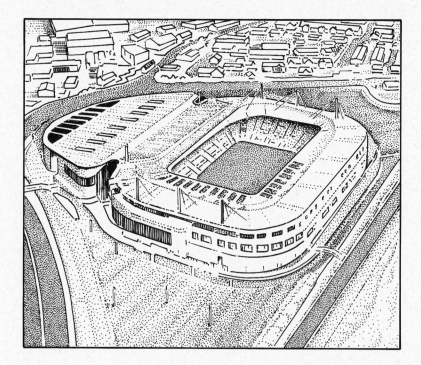

Marks

1. (continued)

(f)   Look at Reference Diagrams Q1B and Q1C on *Page six*.

A new football ground was built for Coventry City Football Club on the site of a former gas works at 343835.

Do you think this change in land use will have brought benefits to the area?

Using map evidence, give reasons for your answer.

_____

_____

_____

_____

_____

_____   **3**

[Turn over

DO NOT WRITE IN THIS MARGIN

KU	ES

Marks

2. Explain how **either** an oxbow lake **or** a waterfall is formed.

You may use diagrams to illustrate your answer.

4

3. **Reference Diagram Q3A: Synoptic Chart for Bordeaux**
    at noon on 30 March 2007

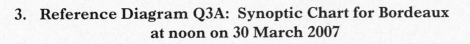

Reference Diagram Q3B

X

Reference Diagram Q3C

Y

Study Reference Diagrams Q3A, Q3B and Q3C.

Which weather station circle, X or Y, is the correct one for Bordeaux at noon on 30 March 2007?

Tick (✓) your choice.

Station circle X ☐   Station circle Y ☐

Give reasons for your choice.

Marks

KU	ES

4

Marks

4.   **Reference Diagram Q4A: Climate Statistics for a Selected Area**

	J	F	M	A	M	J	J	A	S	O	N	D
Temperature (°C)	12	13	15	19	23	25	25	24	17	16	13	12
Rainfall (mm)	66	58	57	57	40	30	20	26	46	68	80	70

Reference Diagram Q4B: Climate Graph for a Selected Area

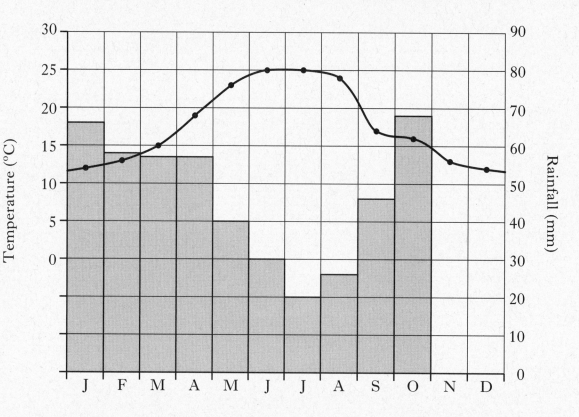

(a)   Look at the table (Reference Diagram Q4A) at the top of the page.
Use this data to complete the climate graph.

2

(b)   Describe, **in detail**, the climate shown in Reference Diagram Q4B.

_____

_____

_____

_____

_____

4

Marks

5.   **Reference Diagram Q5: Desertification**

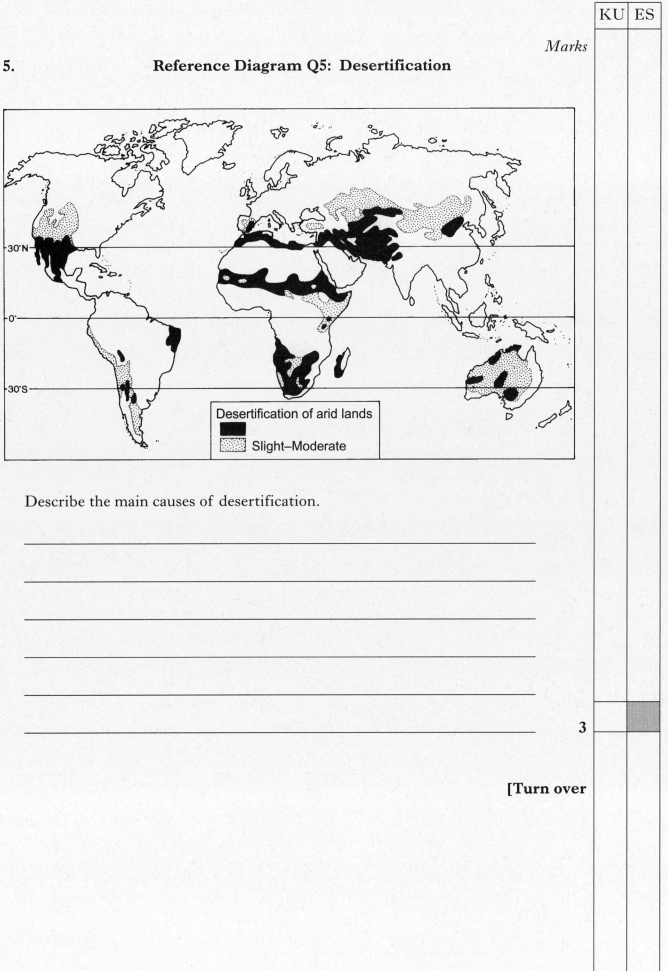

Describe the main causes of desertification.

_____

_____

_____

_____

_____

_____

3

[Turn over

6.   **Reference Diagram Q6A: Location of Prologis Business Park, Coventry**

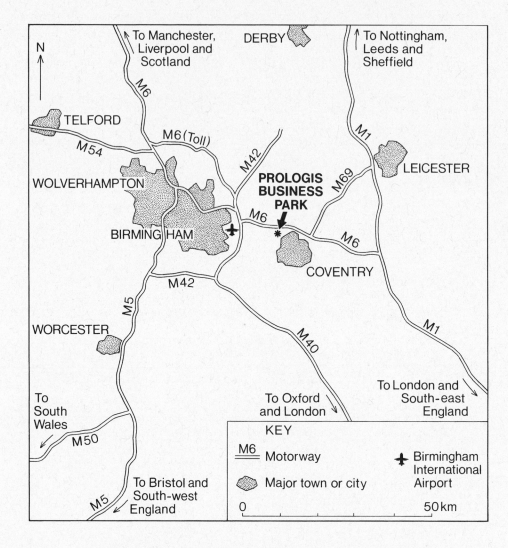

Reference Diagram Q6B: Business Park

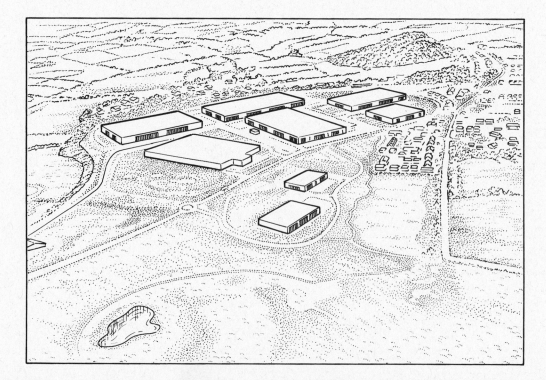

Marks

6. (continued)

(a) Study Reference Diagrams Q6A and Q6B.

Why is this a good location for a business park?

_____

_____

_____

_____

_____

_____

4

(b) Identify **two** gathering techniques which a group of pupils could use to find out how the Business Park affects the area around Coventry.

Give reasons for your answer.

Technique 1 _____

_____

Reason(s) _____

_____

_____

Technique 2 _____

_____

Reason(s) _____

_____

_____

4

[Turn over

7. **Reference Diagram Q7: Solutions to Traffic Congestion in
a British City Centre**

Congested Traffic

Marks

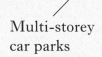

One-way street
system

Ring road

Flexi-time —— SOLUTIONS TO —— Parking restrictions
working   CONGESTION   (eg meters, wardens,
double yellow lines)

Multi-storey   Park and ride
car parks   schemes

Look at Reference Diagram Q7.

Which **two** solutions would be most effective in reducing traffic congestion
in a British city centre?

Give reasons for your choices.

Solution 1 _____

Reasons _____

_____

_____

Solution 2 _____

Reasons _____

_____

_____

4

KU | ES

Marks

8.   **Reference Diagram Q8: Features of the European Union (EU)**

trade barriers with
non members

free trade between
members

currency
(Euros)

economic help to
selected regions

free movement of workers
between member countries

Look at Reference Diagram Q8.

"Membership of the European Union (EU) has both advantages and disadvantages **for the UK**."

What are these advantages and disadvantages?

Advantages _____

_____

_____

_____

Disadvantages _____

_____

_____

_____

4

[Turn over

9. **Reference Diagram Q9A: Population Structure for Hungary in 2009**

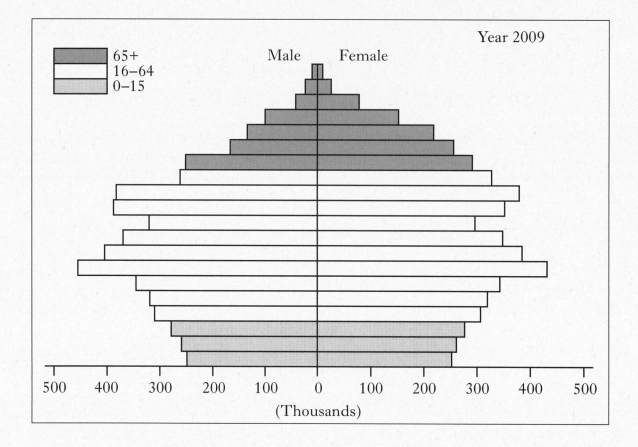

Reference Diagram Q9B: Expected Population Structure for Hungary in 2049

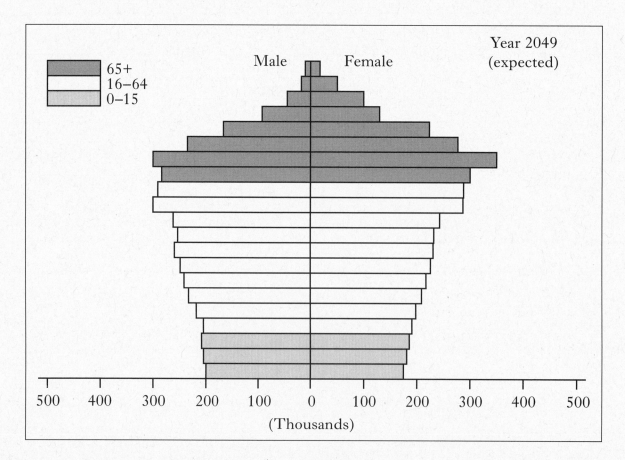

Marks

9. (continued)

Look at Reference Diagrams Q9A and Q9B.

(*a*) Describe the changes which are expected to take place in the population structure of Hungary by 2049.

_____

_____

_____

_____

_____

3

Look at Reference Diagram Q9B.

(*b*) What problems might the expected population structure in 2049 create for Hungary?

_____

_____

_____

_____

_____

_____

4

[Turn over

Marks

10.   **Reference Diagram Q10: United States—Imports and Exports
(2000–2005 Average)**

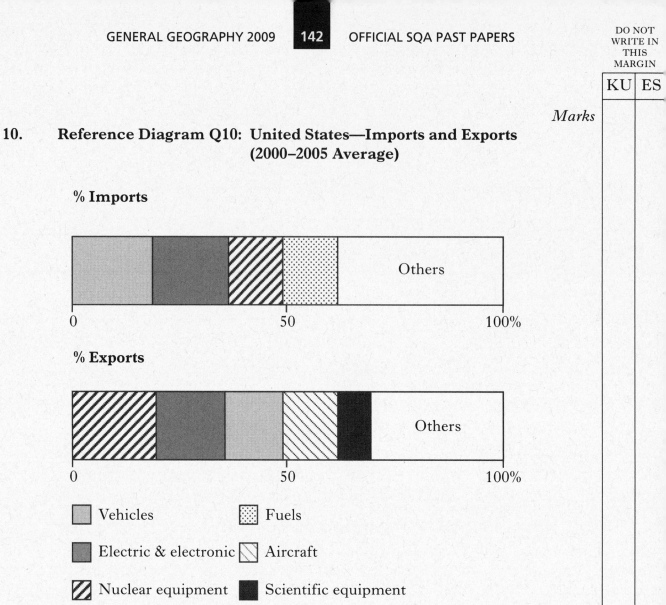

% Imports

% Exports

Vehicles   Fuels

Electric & electronic   Aircraft

Nuclear equipment   Scientific equipment

Look at Reference Diagram Q10.

Give **other** processing techniques which could be used to show the information in the diagrams above.

Why are these techniques suitable?

_____

_____

_____

_____

_____

4

Marks

11. **Reference Diagram Q11: Self-help Schemes**

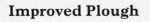

Improved Plough

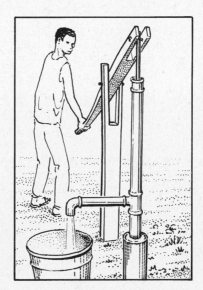

Hand Pump

Stone Lines

Look at Reference Diagram Q11.

Why are self-help schemes suitable for Economically Less Developed Countries (ELDCs)?

_____

_____

_____

_____

_____

3

[*END OF QUESTION PAPER*]

[BLANK PAGE]

[BLANK PAGE]

C

1260/405

NATIONAL QUALIFICATIONS 2009	WEDNESDAY, 27 MAY 1.00 PM – 3.00 PM	**GEOGRAPHY** STANDARD GRADE Credit Level

All questions should be attempted.

Candidates should read the questions carefully.　Answers should be clearly expressed and relevant.

Credit will always be given for appropriate sketch-maps and diagrams.

Write legibly and neatly, and leave a space of about one centimetre between the lines.

All maps and diagrams in this paper have been printed in black only: no other colours have been used.

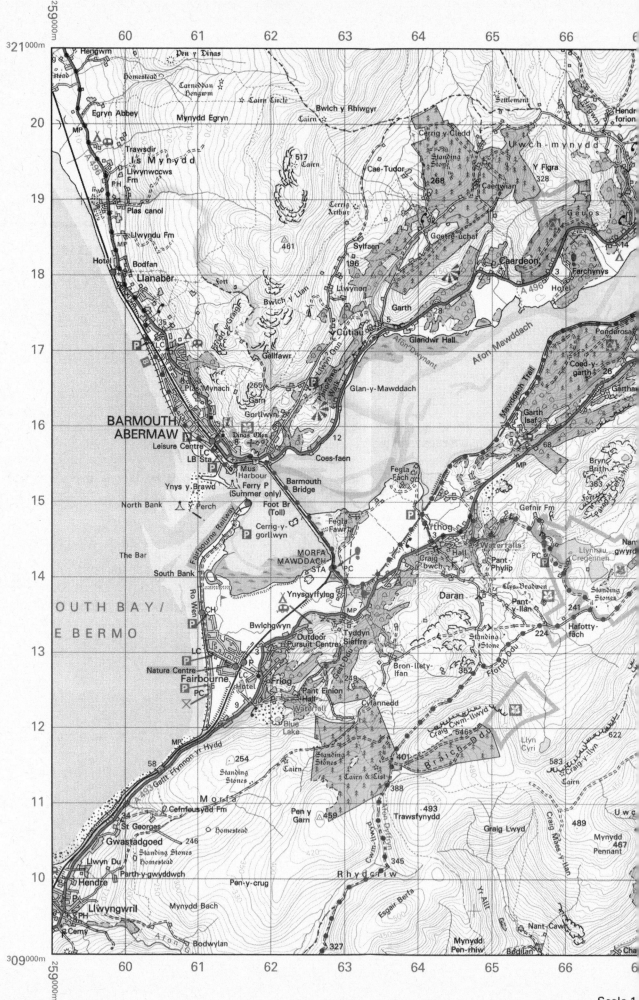

Scale 1

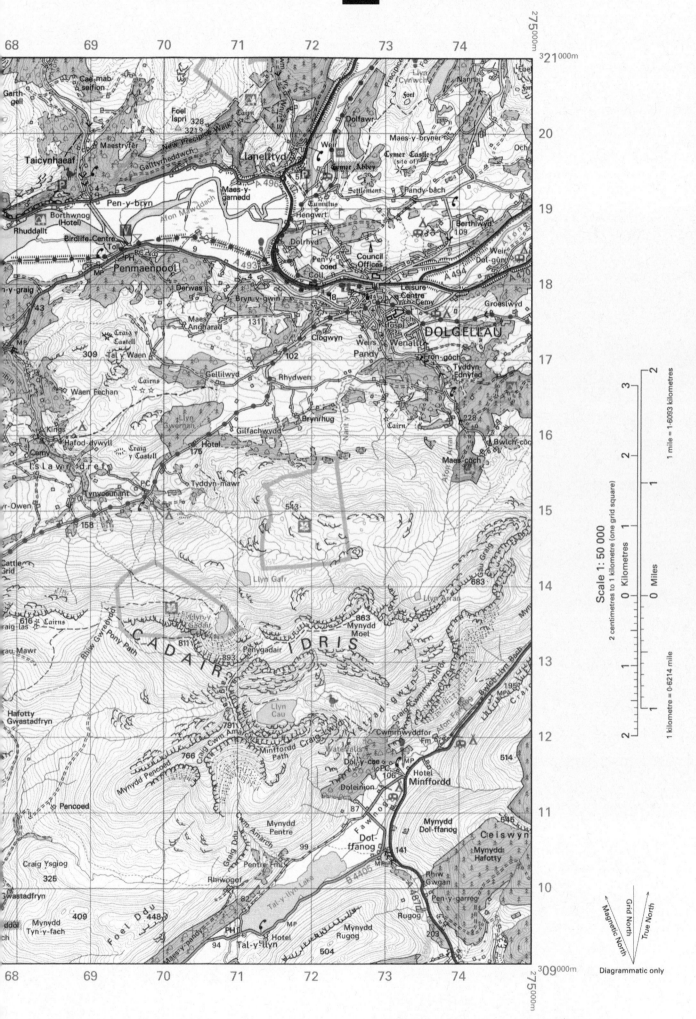

Scale 1: 50 000

2 centimetres to 1 kilometre (one grid square)

1 mile = 1·6093 kilometres

1 kilometre = 0·6214 mile

Ordnance Survey, OS, the OS Symbol and Landranger are registered trademarks of Ordnance Survey, the national mapping agency of Great Britain. Reproduction in whole or in part by any means is prohibited without the prior written permission of Ordnance Survey. For educational use only.

Diagrammatic only

1.   **Reference Diagram Q1A**

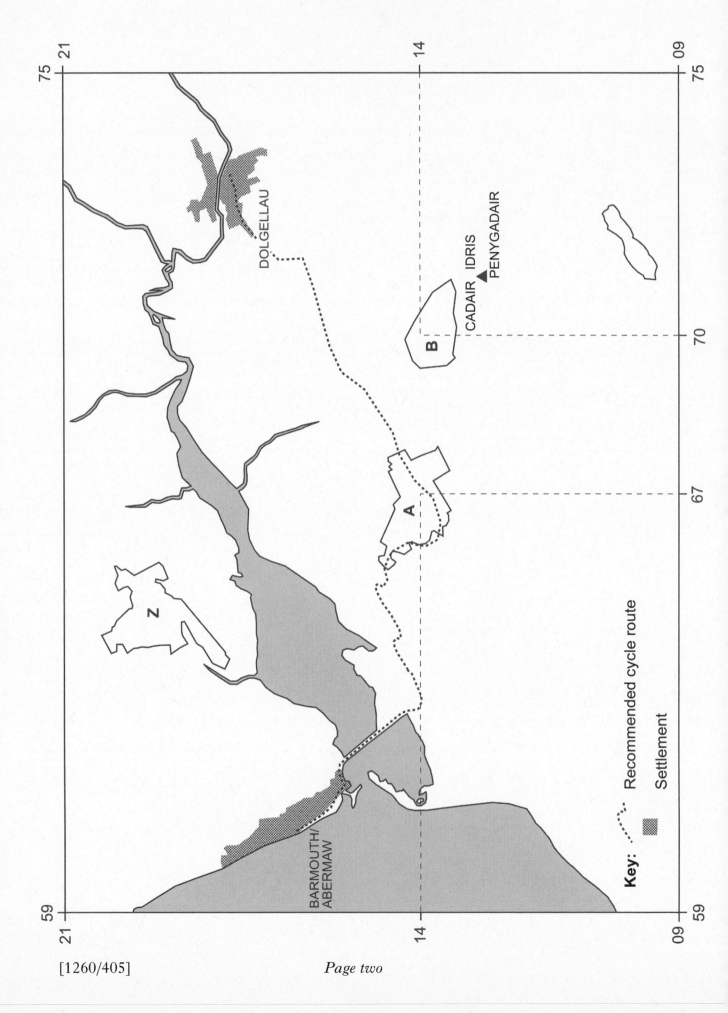

Marks

KU	ES

1. (continued)

This question refers to the OS Map Extract (No 1743/124) of the Dolgellau area and Reference Diagram Q1A on *Page two*.

(*a*) (i) Match each of the features named below with the correct grid reference.

Features: **pyramidal peak**; **corrie**; **truncated spur**; **hanging valley**.

Choose from grid references: 711130, 723125, 715123, 733110.   **3**

(ii) **Explain** how **one** of the features listed in (*a*)(i) was formed.

You may use diagrams to illustrate your answer.   **4**

(*b*)   **Reference Diagram Q1B: Selected Settlement Functions**

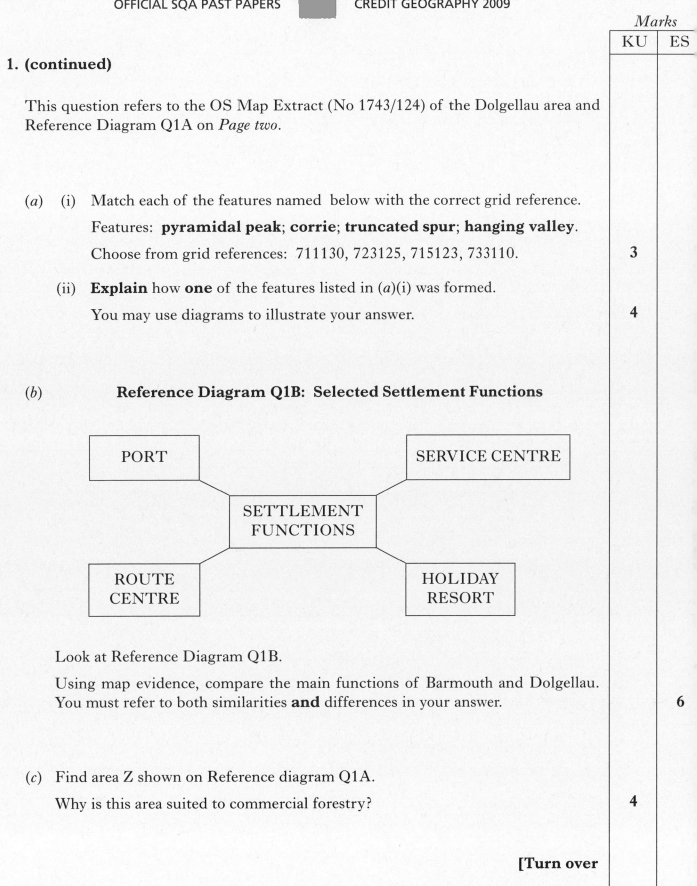

Look at Reference Diagram Q1B.

Using map evidence, compare the main functions of Barmouth and Dolgellau. You must refer to both similarities **and** differences in your answer.   **6**

(*c*) Find area Z shown on Reference diagram Q1A.

Why is this area suited to commercial forestry?   **4**

[Turn over

Mark

KU

1. (continued)

(*d*) A family cycling from Dolgellau to Barmouth have been recommended to follow the route shown on Reference Diagram Q1A.

What are the advantages and disadvantages of taking this route?

(*e*) A group of students wish to carry out a study to compare the two National Trust areas A and B, shown on Reference Diagram Q1A.

Describe gathering techniques which could be used to obtain relevant information.

Give reasons for your choice of techniques.

2.   **Reference Diagram Q2: A V-shaped Valley**

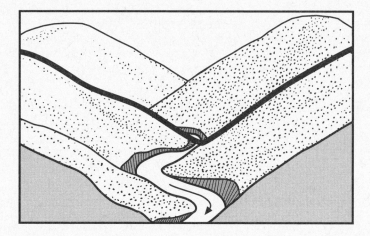

Study Reference Diagram Q2 above.

Explain the formation of a V-shaped valley.

You may use diagrams to illustrate your answer.

4

3.   **Reference Diagram Q3: Synoptic Chart for 15th February 2007**

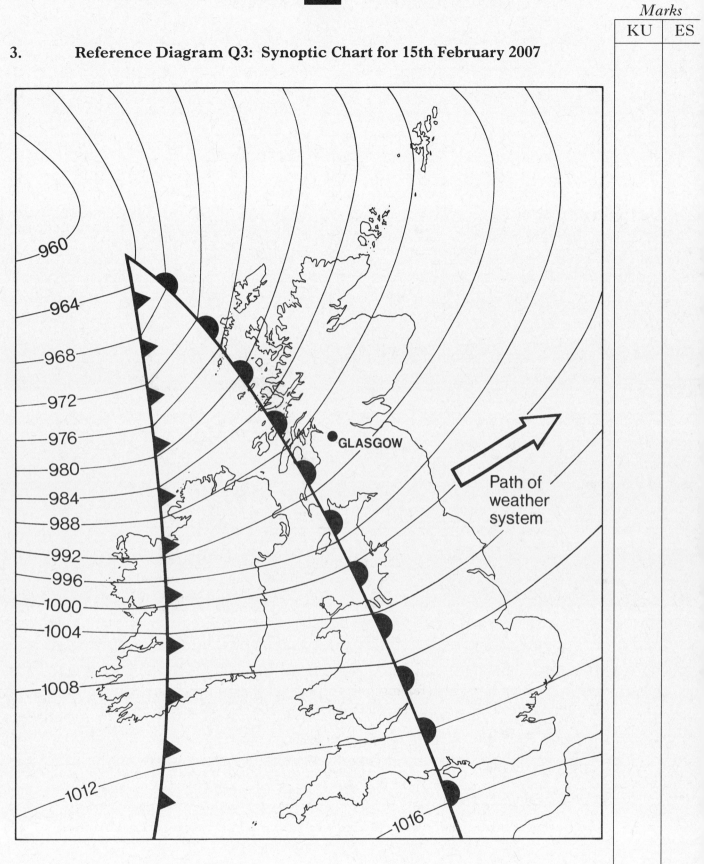

Study Reference Diagram Q3.

Explain the changes which will take place in the weather in Glasgow over the next 24 hours.

6

[Turn over

4.

Reference Diagram Q4

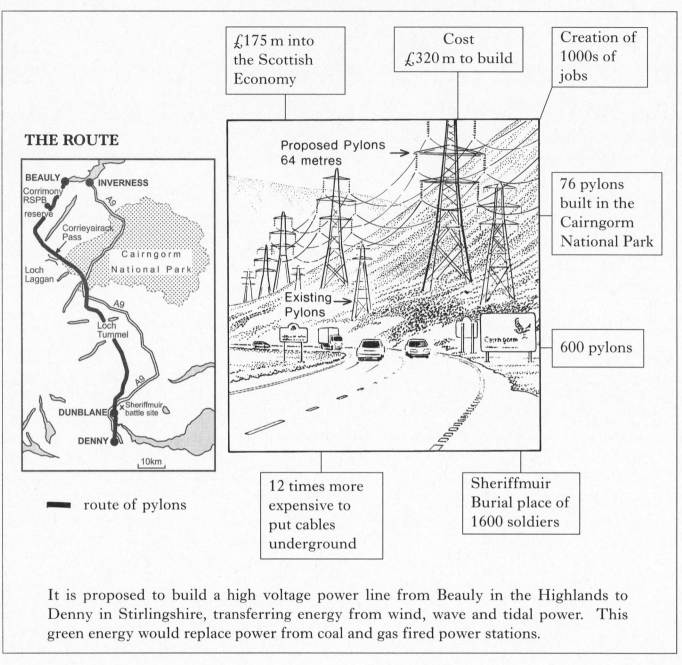

£175 m into the Scottish Economy

Cost £320 m to build

Creation of 1000s of jobs

76 pylons built in the Cairngorm National Park

600 pylons

12 times more expensive to put cables underground

Sheriffmuir Burial place of 1600 soldiers

THE ROUTE

— route of pylons

It is proposed to build a high voltage power line from Beauly in the Highlands to Denny in Stirlingshire, transferring energy from wind, wave and tidal power. This green energy would replace power from coal and gas fired power stations.

Mark

KU

Study Reference Diagram Q4.

"The benefits brought about by the power line will be more important than the damage to the countryside."

Do you agree fully with this statement?

Give reasons for your answer.

5.   **Reference Diagram Q5: Land Use in a Scottish City**

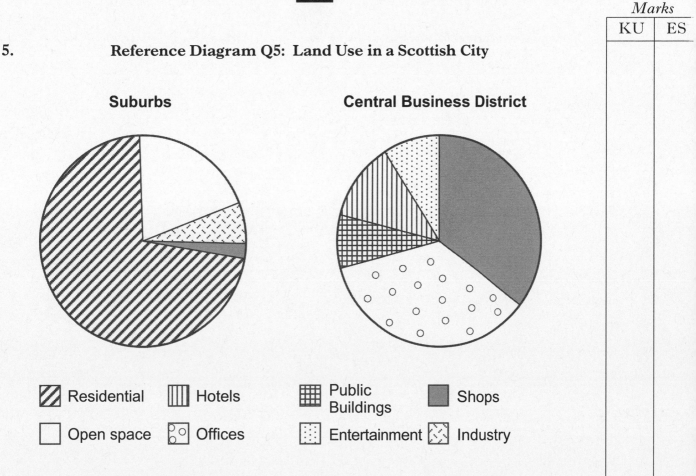

Look at Reference Diagram Q5.

Give reasons for the patterns of the land use shown in both the suburbs and the Central Business District.

6

[Turn over

6.   **Reference Diagram Q6A: Farming Landscape in 1950**

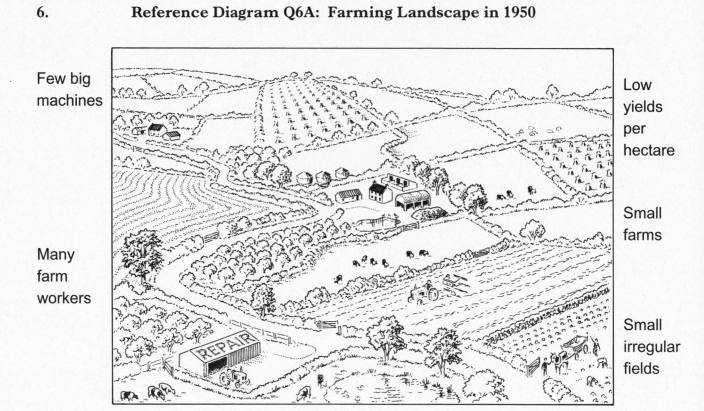

Few big machines

Low yields per hectare

Small farms

Many farm workers

Small irregular fields

Reference Diagram Q6B: Farming Landscape in 2005

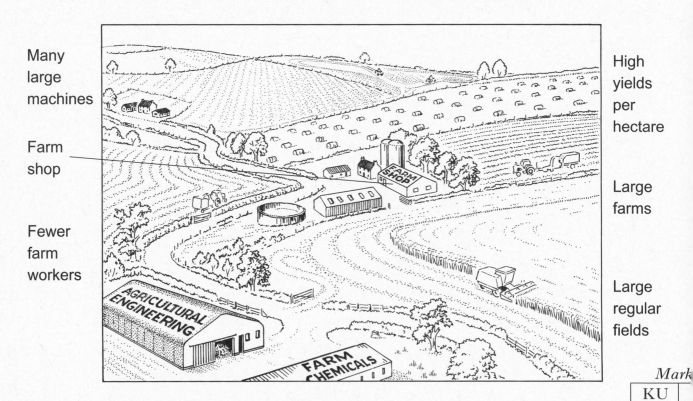

Many large machines

High yields per hectare

Farm shop

Large farms

Fewer farm workers

Large regular fields

Mark

KU

Look at Reference Diagrams Q6A and Q6B.

"The changes in farming since 1950 have brought more benefits than problems."

Do you agree fully with this statement?

Give reasons for your answer.

7.

Reference Diagram Q7A: Factors influencing the Location of a Science Park

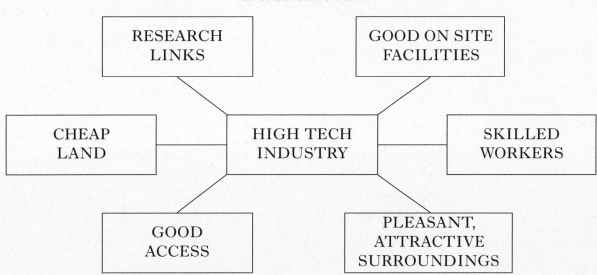

Reference Map Q7B: Location of University of Southampton Science Park

Reference Diagram Q7C: Site of University of Southampton Science Park

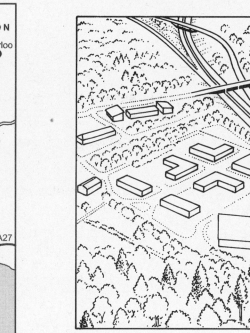

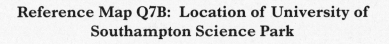

═══ Motorway   ⊕ Airport

━━━ A class road

↟ Ferry route

─── Railway

● Railway station ▩ Built up area

	Marks	
	KU	ES

Look at Reference Diagrams Q7A, Q7B and Q7C.

In your opinion, which **three** factors were most important in the location of the University of Southampton Science Park?

Give reasons to support your choice.

5

Marks

KU

8.   **Reference Diagram Q8A: Population Distribution Map of Europe**

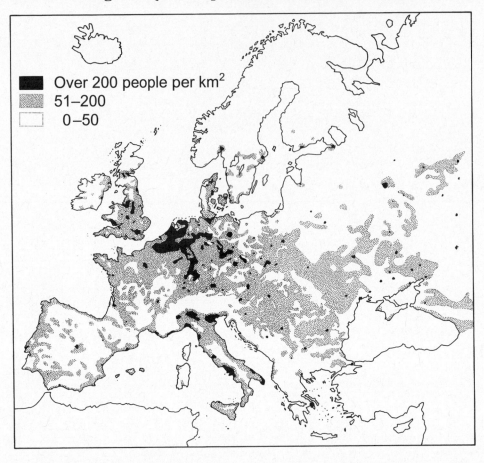

Reference Diagram Q8B: Relief and International Boundaries of Europe

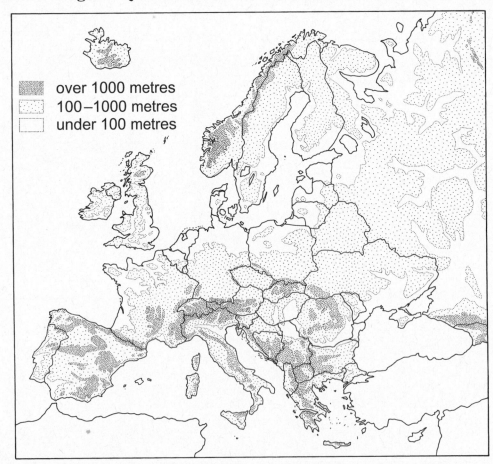

Study Reference Diagrams Q8A and Q8B.

Describe in detail the population distribution in Europe.

9.   **Reference Diagram Q9A: Change in worldwide rural and urban populations**

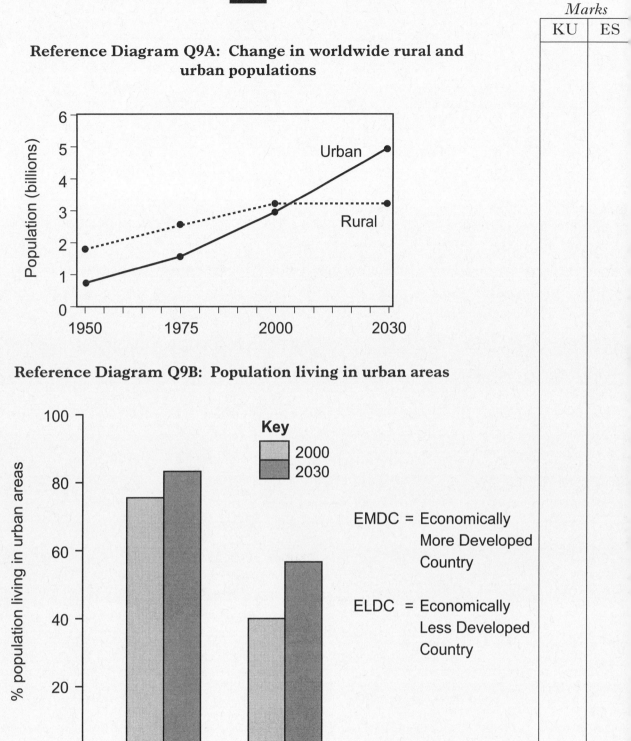

Reference Diagram Q9B: Population living in urban areas

Look at Reference Diagrams Q9A and Q9B.

(*a*) Describe the trends shown in both graphs.

(*b*) **Explain** why urban populations are expected to continue increasing.

4

5

[**Turn over**

Mar

KU

10.   **Reference Diagram Q10: Percentages of Population aged over 65**

Country	1960	1970	1980	1990	2000	2010
Japan	6·0	7·0	9·0	12·0	16·0	22·0
Nigeria	2·3	2·4	2·5	2·5	2·7	3·2

Study Reference Diagram Q10.

What other techniques could be used to show the information in the table?

Give reasons for your choices.

11.   **Reference Diagram Q11: Different Types of Aid in ELDCs**

Free Gift   Small Scale

Experienced
Field   CHARITY — Depends on
Workers   AID   Donations

No Loans   No Conditions
or Debts

Tied
Aid

Large   Project
Loans   specific

BILATERAL
AID

Increased   Funding for
Trade   Large
        Links with   Projects
        Donor
        Country

"Aid from charities is better than bilateral aid."

Do you agree fully with this statement?

Give reasons for your answer.

[END OF QUESTION PAPER]

[BLANK PAGE]

FOR OFFICIAL USE

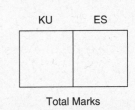

KU | ES

Total Marks

1260/403

NATIONAL
QUALIFICATIONS
2010

THURSDAY, 6 MAY
10.25 AM–11.50 AM

GEOGRAPHY
STANDARD GRADE
General Level

Fill in these boxes and read what is printed below.

Full name of centre

Town

Forename(s)

Surname

Date of birth

Day   Month   Year   Scottish candidate number   Number of seat

1 Read the whole of each question carefully before you answer it.

2 Write in the spaces provided.

3 Where boxes like this ☐ are provided, put a tick ✓ in the box beside the answer you think is correct.

4 Try all the questions.

5 Do not give up the first time you get stuck: you may be able to answer later questions.

6 Extra paper may be obtained from the Invigilator, if required.

7 Before leaving the examination room you must give this book to the Invigilator. If you do not, you may lose all the marks for this paper.

Extract No 1785/90

1:50 000 Scale
Landranger Series

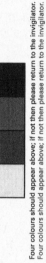

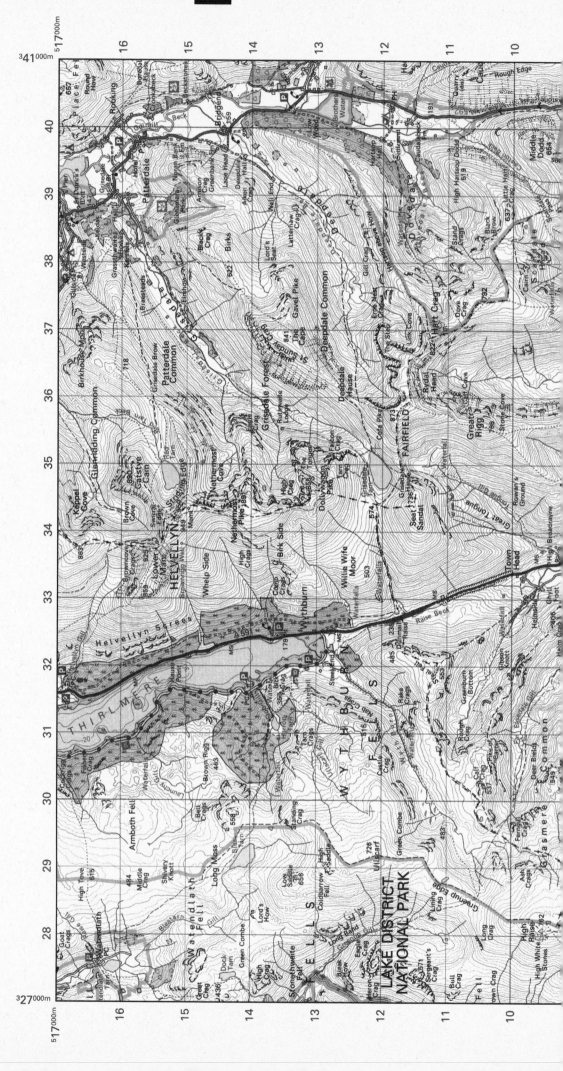

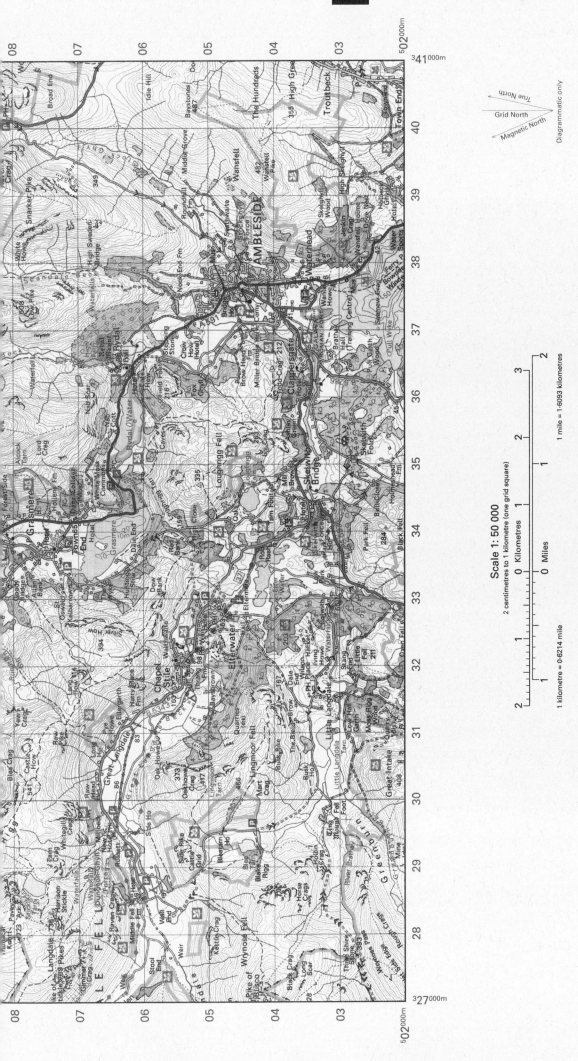

Scale 1: 50 000

2 centimetres to 1 kilometre (one grid square)

1 kilometre = 0·6214 mile

1 mile = 1·6093 kilometres

True North
Grid North
Magnetic North

Diagrammatic only

Extract produced by Ordnance Survey 2009. Licence: 100035658

1.

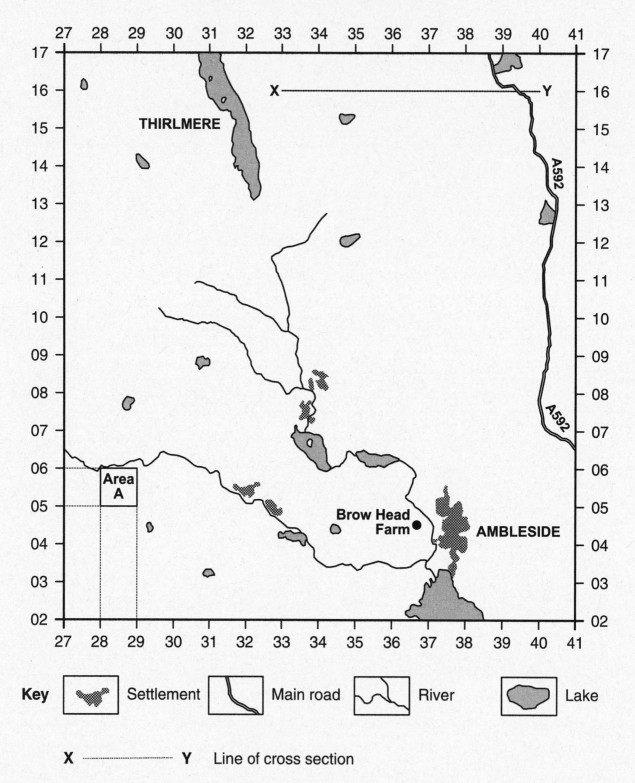

Diagram Q1A: The Ambleside Area

Key
Settlement
Main road
River
Lake

X ·············· Y Line of cross section

Marks

1. (continued)

Look at the Ordnance Survey Map Extract (No 1785/90) of the Ambleside area and Diagram Q1A on *Page two*.

Diagram Q1B: Cross Section from X (330160) to Y (400160)

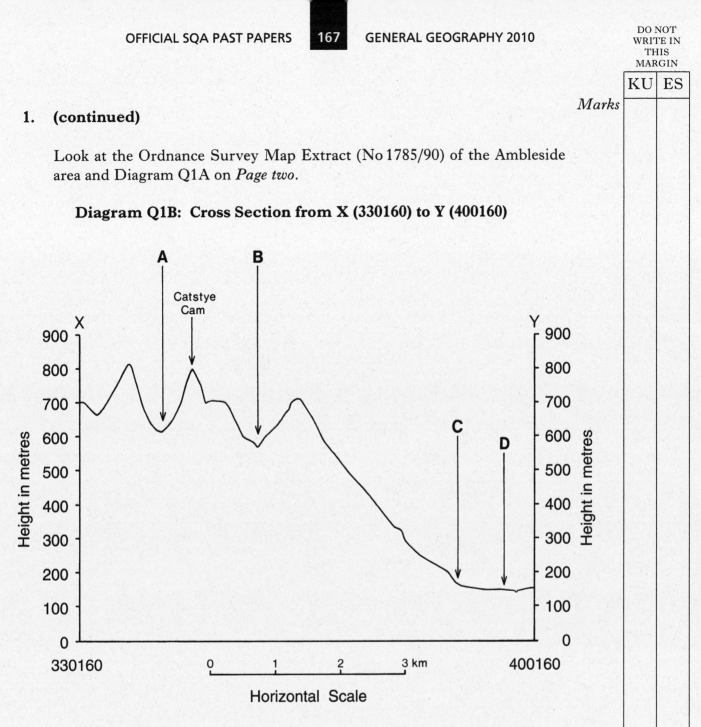

(a) Look at Diagram Q1B and Diagram Q1A. Find cross section XY on the map extract.

Match the features A, B, C and D on the cross section XY with the correct description in the table below.

Feature	Letter
A592	
Brown Cove	
Mixed wood	
Red Tarn Beck	

3

Marks

1. (continued)

(*b*) (i) Match the glacial features in the table with the grid references below.

2807   3115   3108   3006

Glacial Feature	Grid Reference
Hanging valley	
Corrie with tarn	
Misfit stream	
Ribbon lake	

3

(ii) **Explain** how **one** of the glacial features named in the table above was formed.

You may use a diagram(s) to illustrate your answer.

_____

_____

_____

_____

_____

_____

_____

_____

_____

3

Marks

1. (continued)

Diagram Q1C: A Holiday Home in the Lake District

Look at Diagrams Q1A and Q1C.

(c) There are plans to build twenty holiday homes in Area A (grid square 2805). Each home will look like the one shown in Diagram Q1C above.

Do you agree fully that these plans should go ahead?

Using map evidence, give reasons for your answer.

_____

_____

_____

_____

_____

_____

4

[Turn over

Marks

1.   (continued)

(*d*) Do you think it will be possible for Ambleside (3704) to grow much further?

Give map evidence to support your answer.

_____

_____

_____

_____

_____

_____

4

(*e*) **"Brow Head Farm in grid square 3604 is likely to be a mixed farm."**

Do you agree fully with this statement?

Give reasons for your answer.

_____

_____

_____

_____

_____

3

KU	ES

Marks

1.   (continued)

Diagram Q1D: Single Carriageway Road   **Diagram Q1E: Dual Carriageway**

Look at Diagrams Q1A, Q1D and Q1E.

(*f*)   There is a plan to widen all of the A592 to a dual carriageway.

Why might people object to this?

Give map evidence to support your answer.

_____

_____

_____

_____

_____

_____   **4**

[Turn over

2.   **Diagram Q2: Synoptic Chart, 6 January, 0600 hours**

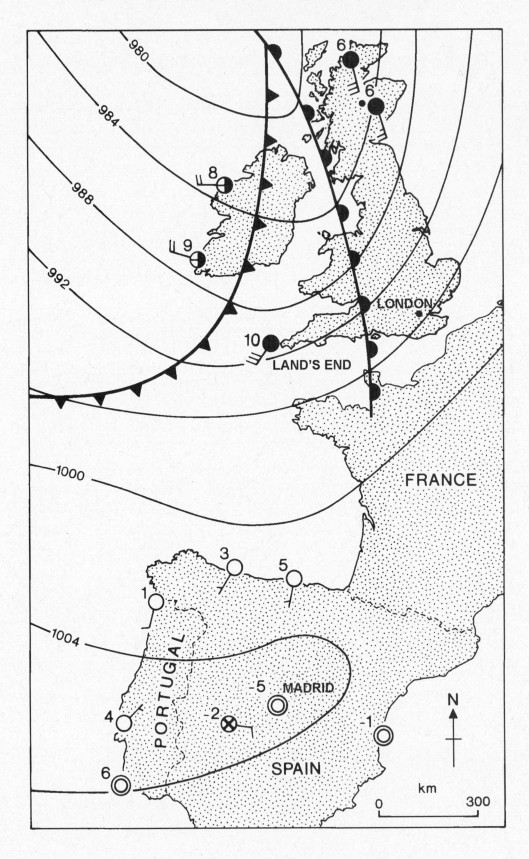

2.   (continued)

Study Diagram Q2.

(a) Complete the station circle below to show the weather conditions at **London** on Diagram Q2.

Weather conditions at London

Wind from South West
Cloud cover: 4 oktas
Rain: Drizzle
Wind speed: 15 knots

8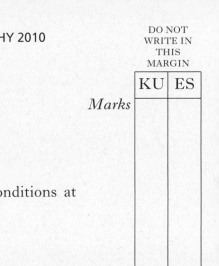

3

(b) Study Diagram Q2.

Match the weather system to the locations given in the table.

Choose from: Anticyclone   Depression

Location	Weather System
British Isles	
Spain	

Give reasons for your answer.

_____

_____

_____

_____

_____

4

[Turn over

3.

Diagram Q3A: Selected Climate Regions

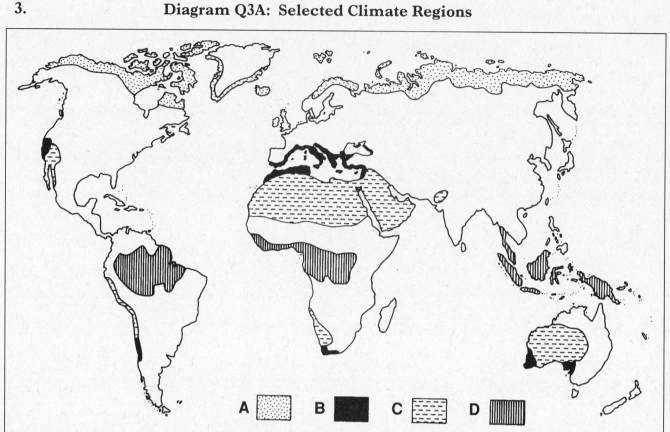

Diagram Q3B: Selected Climate Graphs

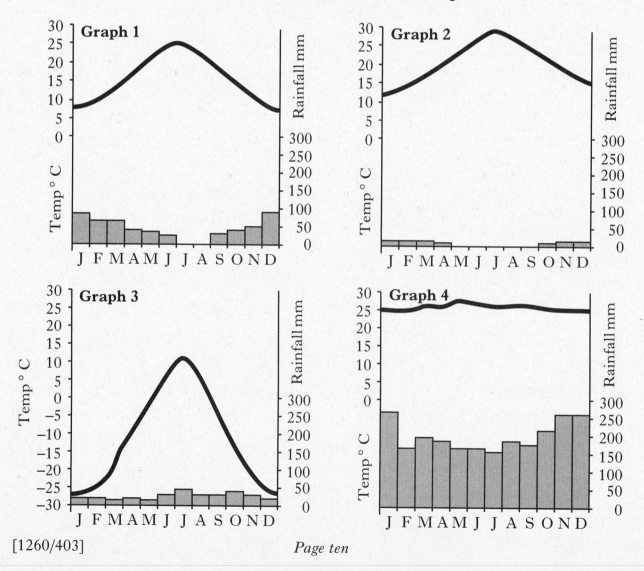

Marks

3.   **(continued)**

Look at Diagrams Q3A and Q3B.

(a) Complete the table below by adding the appropriate letter or number.

Climate	Map Area	Graph
Hot desert		2
Equatorial Rainforest	D	
Mediterranean		1
Tundra	A	

2

(b) Describe, **in detail**, the main features of the climate shown in **Graph 4**.

_____

_____

_____

_____

_____

_____

3

[Turn over

4.   **Diagram Q4: New Developments at the Edge of Glasgow**

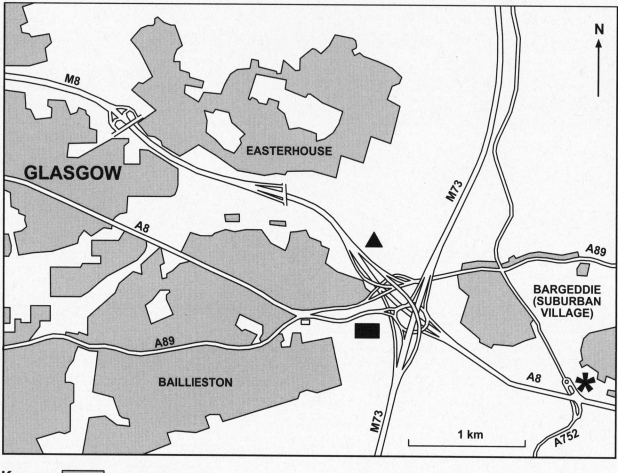

Key

Built-up area

New Developments

  Retail Park

  Supermarket

  Leisure Complex

Marks

4. (continued)

Look at Diagram Q4.

(a) Select **one** of the developments shown and describe the advantages **and** disadvantages of its location.

Selected development _____

Advantages _____

_____

_____

Disadvantages _____

_____

_____

4

(b) What **two** techniques could local pupils use to gather information about land use changes in the area shown in Diagram Q4?

Give reasons for your choices.

Technique 1 _____

Reason _____

_____

Technique 2 _____

Reason _____

_____

4

[Turn over

5.   **Diagram Q5A: Aerial View in 2005 of Site for London Olympics**

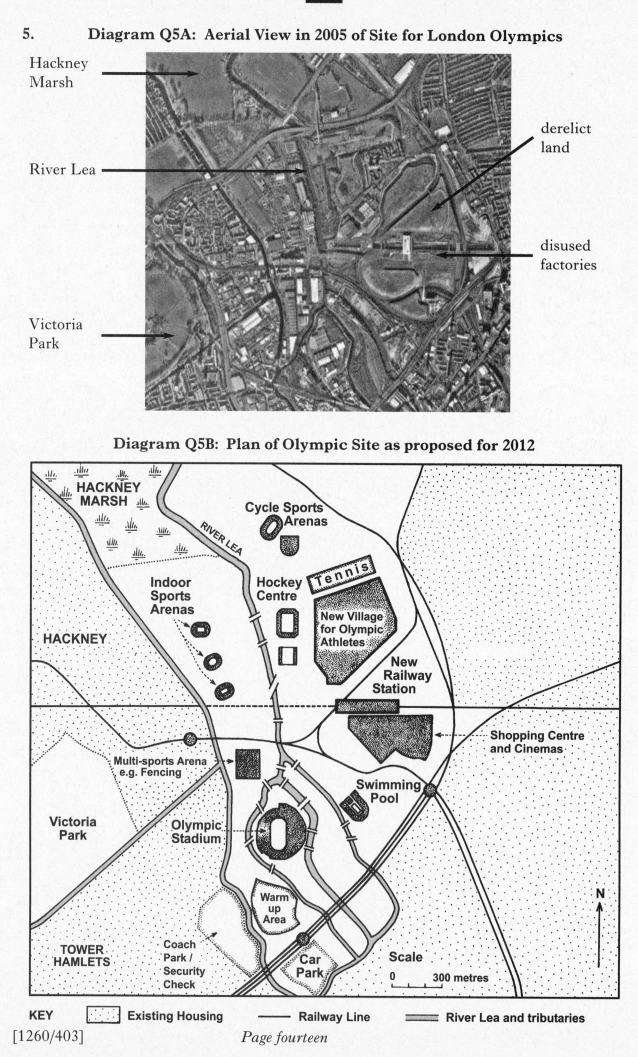

Diagram Q5B: Plan of Olympic Site as proposed for 2012

DO NOT WRITE IN THIS MARGIN

KU	ES

Marks

5. (continued)

Study Diagrams Q5A and Q5B.

The Olympic Games will be held in London in 2012.

How is this likely to benefit London?

4

[Turn over

6.   **Diagram Q6A: Braemore Farm, Northwest Scotland, in 1980**

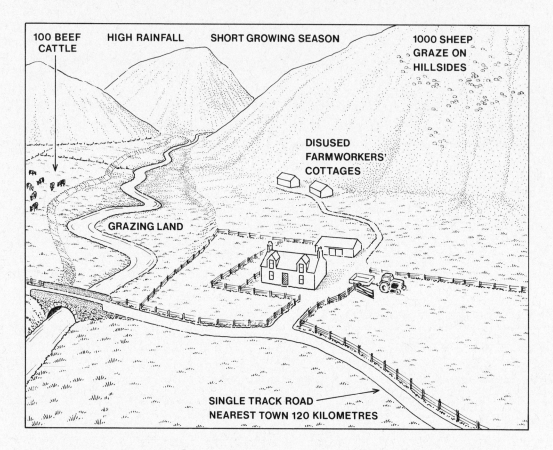

Diagram Q6B: Braemore Farm, Northwest Scotland, in 2010

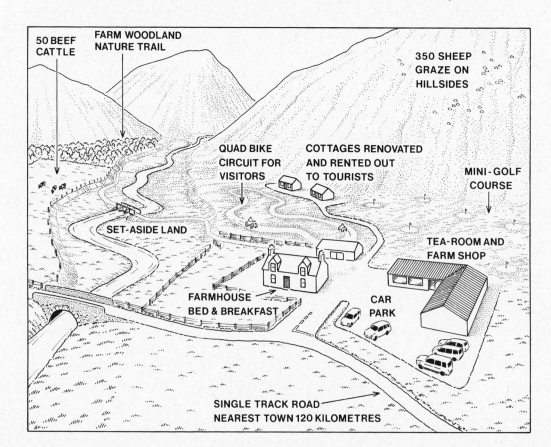

Marks

6. **(continued)**

Study Diagrams Q6A and Q6B.

Braemore Farm in northwest Scotland has changed since 1980. Are these changes likely to have improved the farm?

Give reasons for your answer.

_____

_____

_____

_____

_____

4

[Turn over

KU | ES

Marks

7.   **Diagram Q7A: Age Structure of Population for Botswana (2000)**

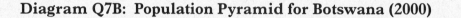

Age Group	Males (% of population)	Females (% of population)
60+	3	3
20–59	20	23
0–19	25	26

Diagram Q7B: Population Pyramid for Botswana (2000)

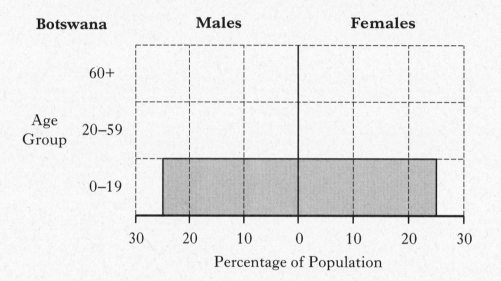

(*a*)   Use the information shown in Diagram Q7A to complete the population pyramid for Botswana above (Diagram Q7B).

3

7. (continued)

Diagram Q7C: Life Expectancy (2007)

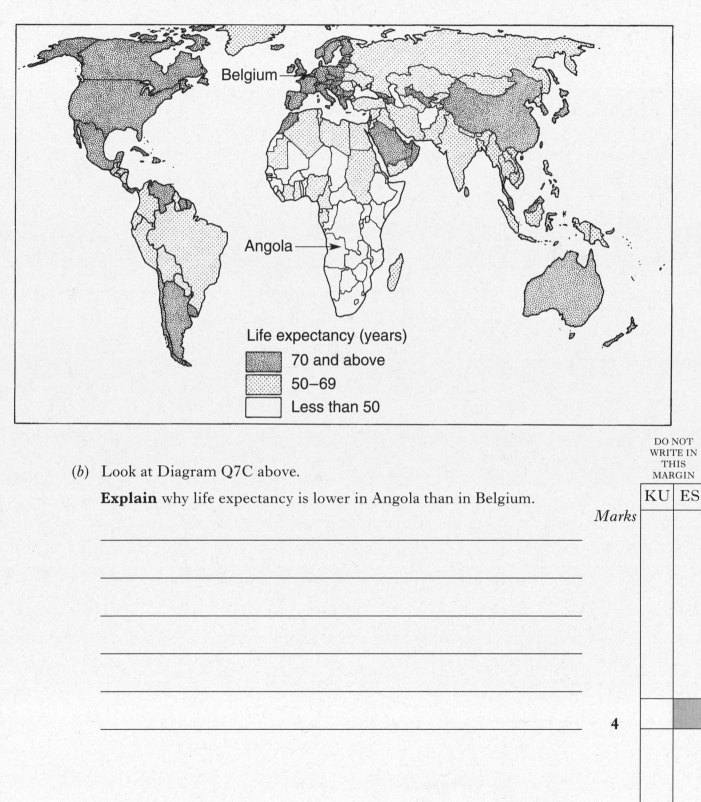

(b) Look at Diagram Q7C above.

Explain why life expectancy is lower in Angola than in Belgium.

Marks

4

[Turn over

Marks

8.

Diagram Q8: Rural/Urban Population in Selected Countries

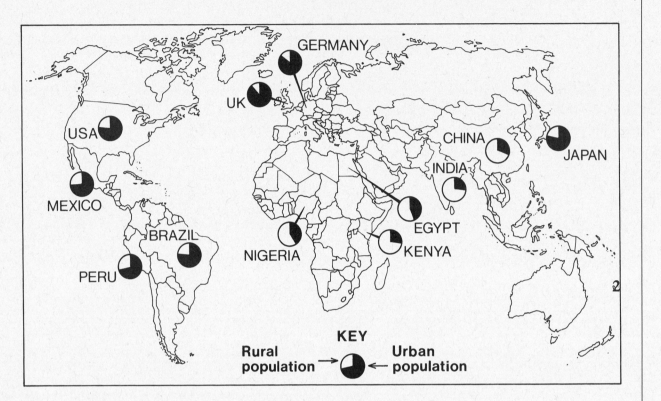

Look at Diagram Q8.

Name **two** other techniques which could be used to show the population information in Diagram Q8.

Give reasons for your choices.

Technique 1 _____

Reason(s) _____

_____

_____

Technique 2 _____

Reason(s) _____

_____

_____

4

Marks

9. Developing countries pay a tax (tariff) to sell their goods in EU countries.

    What problems can this cause for these developing countries?

    _____

    _____

    _____

    _____

    _____

    _____   3

[Turn over

10.

Diagram Q10A: Haiti

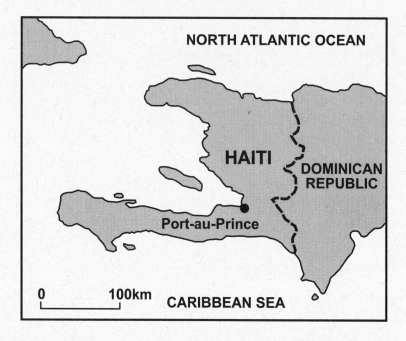

Diagram Q10B: Effects of Hurricanes in Haiti, August/September 2008

Deaths	500
Homeless	1 million
Crops destroyed	coffee, tobacco and sugar
Schools destroyed	30
Estimated cost of rebuilding	$180 million

Diagram Q10C: Types of Aid

Short-term Aid	Long-term Aid
Clean water	Rebuilding homes
Food	Road building
Emergency shelter	Electricity network
Medicines	Building hospitals

Marks

10. (continued)

Look at Diagrams Q10A, Q10B and Q10C.

In 2008, four hurricanes hit Haiti in twenty one days causing widespread destruction and flooding.

Which type of aid, **short-term** or **long-term**, would have been more useful to Haiti?

Explain your answer **in detail**.

_____

_____

_____

_____

_____

_____ **4**

[END OF QUESTION PAPER]

[BLANK PAGE]

2010

[BLANK PAGE]

C

1260/405

| NATIONAL QUALIFICATIONS 2010 | THURSDAY, 6 MAY 1.00 PM – 3.00 PM | GEOGRAPHY STANDARD GRADE Credit Level |

All questions should be attempted.

Candidates should read the questions carefully. Answers should be clearly expressed and relevant.

Credit will always be given for appropriate sketch-maps and diagrams.

Write legibly and neatly, and leave a space of about one centimetre between the lines.

All maps and diagrams in this paper have been printed in black only: no other colours have been used.

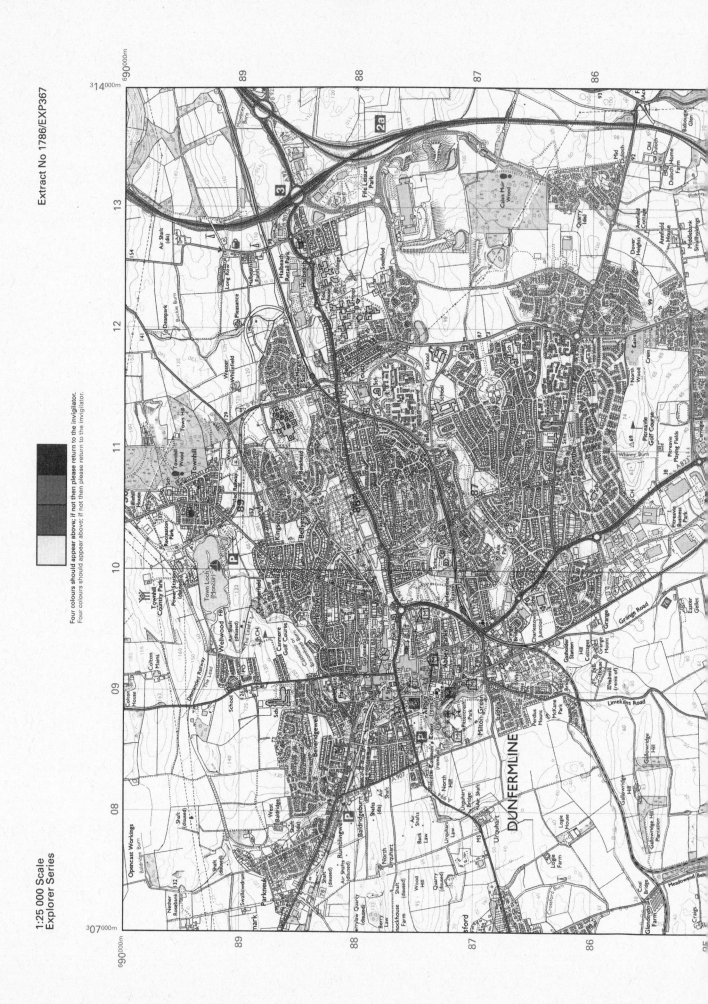

Extract No 1786/EXP367

1:25 000 Scale
Explorer Series

Four colours should appear above; if not then please return to the invigilator.
Four colours should appear above; if not then please return to the invigilator.

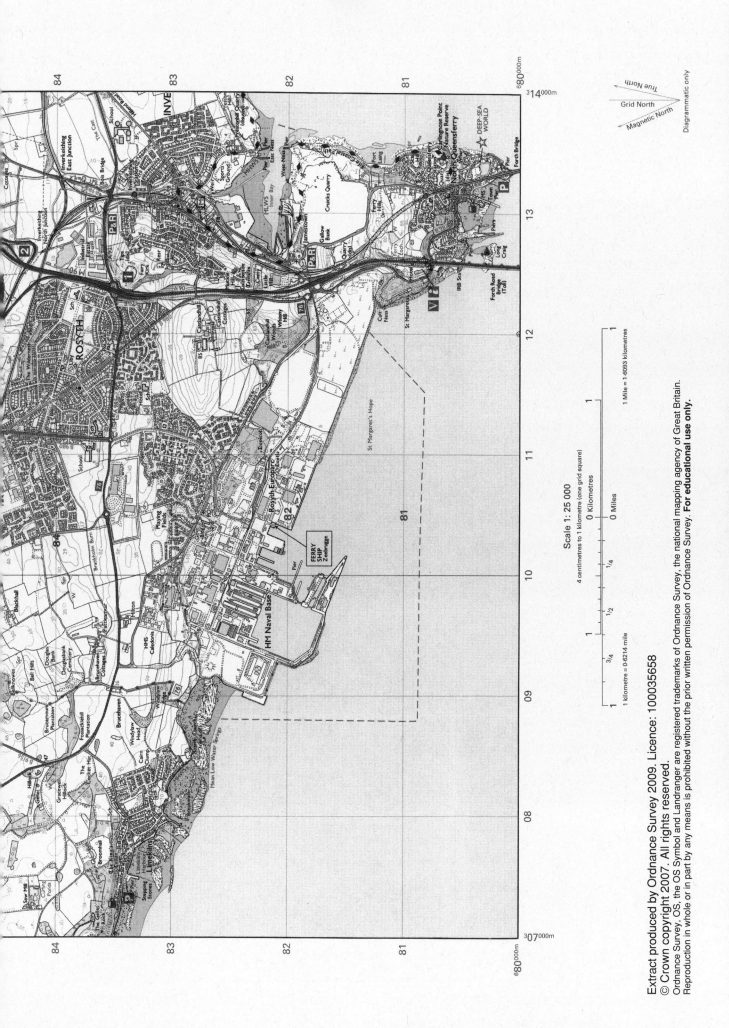

Scale 1:25 000

4 centimetres to 1 kilometre (one grid square)

1 Mile = 1·6093 kilometres

1.   **Diagram Q1**

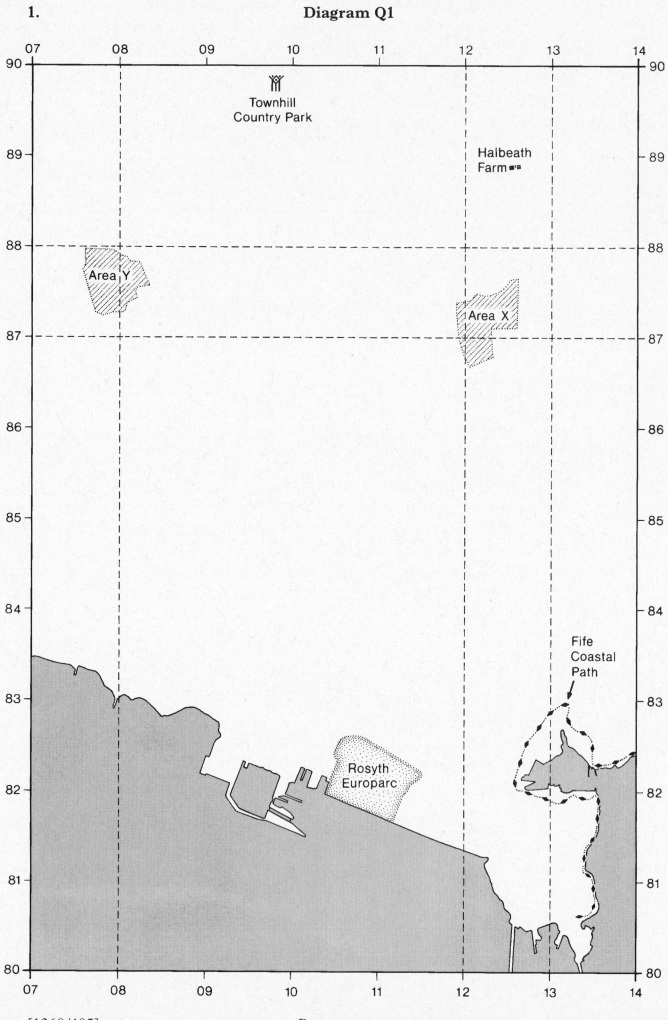

Marks

KU | ES

1. (continued)

This question refers to the OS Map (No 1786/EXP367) of the Dunfermline area and to Diagram Q1 on *Page two*.

(*a*) Find the route of the Fife Coastal Path between 134805 and 140824.

Describe the advantages **and** disadvantages of this route for walkers.

5

(*b*) What techniques could a group of Geography students use to gather information about Townhill Country Park (0989)?

Give reasons for your choice of techniques.

5

(*c*) What is the main present day function of Dunfermline?

Choose from: service centre   tourist centre.

Use map evidence to support your answer.

4

(*d*) There is a plan to build a new housing estate at either Area X (around 120870) or Area Y (around 082877).

Which location is better?

Give reasons for your choice.

5

(*e*) **"Halbeath Farm at 126889 is an excellent location for farming."**

Do you agree fully with this statement?

Give reasons for your answer.

4

(*f*) Rosyth Europarc (110820) is an industrial estate.

Explain the advantages of its location.

4

[Turn over

2.   **Diagram Q2A: A Meander**   **Diagram Q2B:
                                                    Cross Section of a Meander**

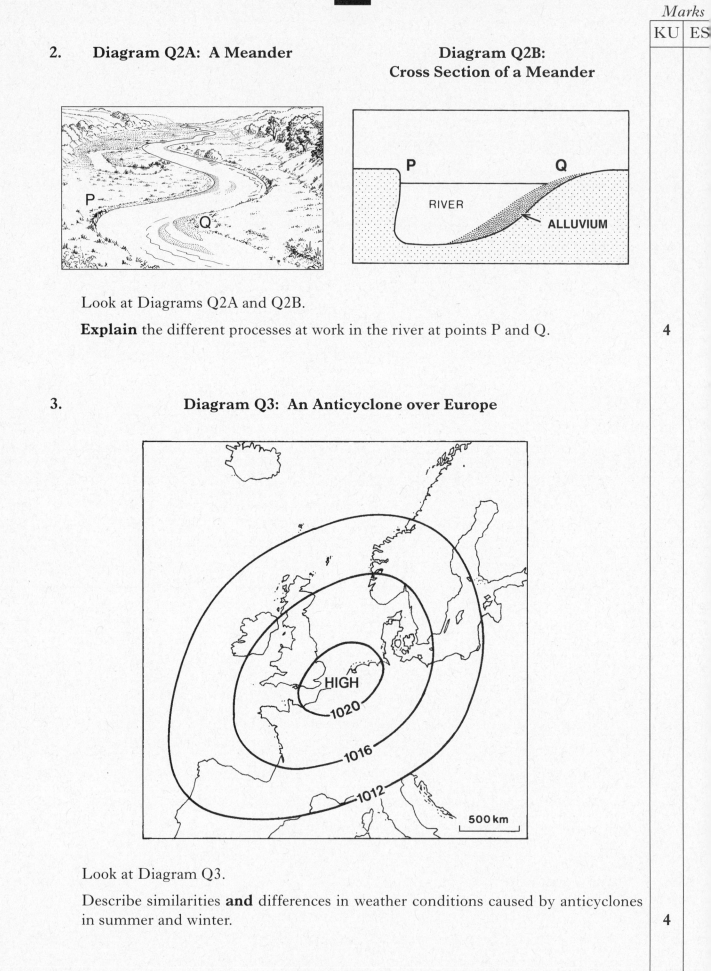

Look at Diagrams Q2A and Q2B.

Explain the different processes at work in the river at points P and Q.   4

3.   **Diagram Q3: An Anticyclone over Europe**

Look at Diagram Q3.

Describe similarities **and** differences in weather conditions caused by anticyclones
in summer and winter.   4

Marks
KU | ES

4.

**Diagram Q4A:
The Sahel Zone, Africa**

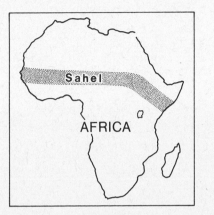

**Diagram Q4B:
Population (millions) in Sahel Countries**

Country	Year		
	1950	1995	2005
Ethiopia	20	52	67
Sudan	9	30	35
Chad	3	6	9

Diagram Q4C: Causes of Desertification

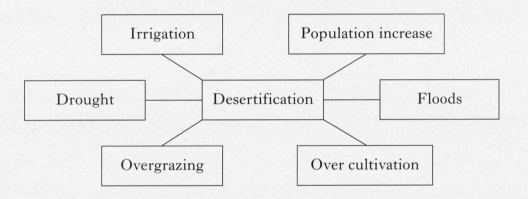

Look at Diagrams Q4A, Q4B and Q4C.

"Population is the main cause of desertification."

Do you agree fully with this statement?

Give reasons for your answer.

5

[Turn over

Marks

KU | ES

5.   **Diagram Q5: Development at Menie Estate**

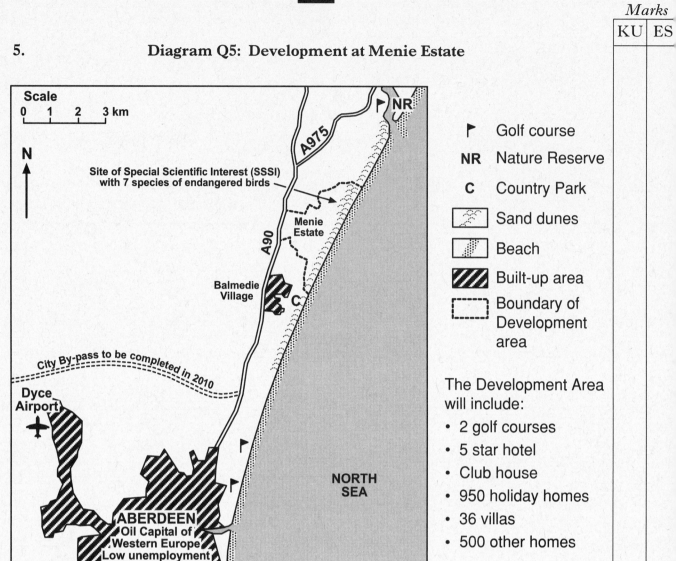

Legend:
- ⚑ Golf course
- **NR** Nature Reserve
- **C** Country Park
- Sand dunes
- Beach
- Built-up area
- Boundary of Development area

The Development Area will include:
- 2 golf courses
- 5 star hotel
- Club house
- 950 holiday homes
- 36 villas
- 500 other homes

Look at Diagram Q5.

A foreign businessman is developing the **Menie Estate** as a luxury golf resort.

What are the advantages **and** disadvantages of this development for the area?   6

6.   **Diagram Q6A: Land Use Zones in Urban Areas**

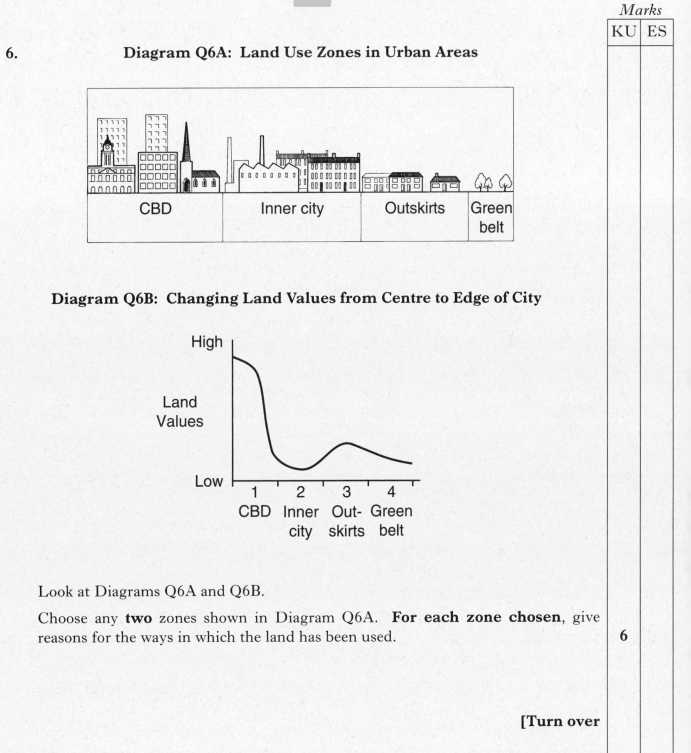

Diagram Q6B: Changing Land Values from Centre to Edge of City

Look at Diagrams Q6A and Q6B.

Choose any **two** zones shown in Diagram Q6A. **For each zone chosen**, give reasons for the ways in which the land has been used.

6

[Turn over

7.   **Diagram Q7: Percentage Income at Redland Farm**

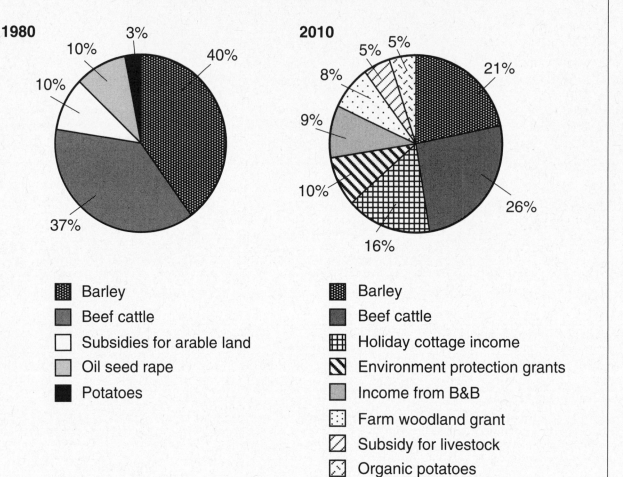

1980

40%
3%
10%
10%
37%

2010

5% 5%
8%
9%
21%
10%
16%
26%

▦ Barley	▦ Barley
◼ Beef cattle	◼ Beef cattle
☐ Subsidies for arable land	▦ Holiday cottage income
◻ Oil seed rape	◨ Environment protection grants
■ Potatoes	◻ Income from B&B
	⬚ Farm woodland grant
	◨ Subsidy for livestock
	⬚ Organic potatoes

Look at Diagram Q7.

The sources of this farmer's income have changed since 1980.

Give reasons for these changes.

6

8.

**Diagram Q8A:
Kenya, Population Density**

**Diagram Q8B:
Kenya, Annual Rainfall**

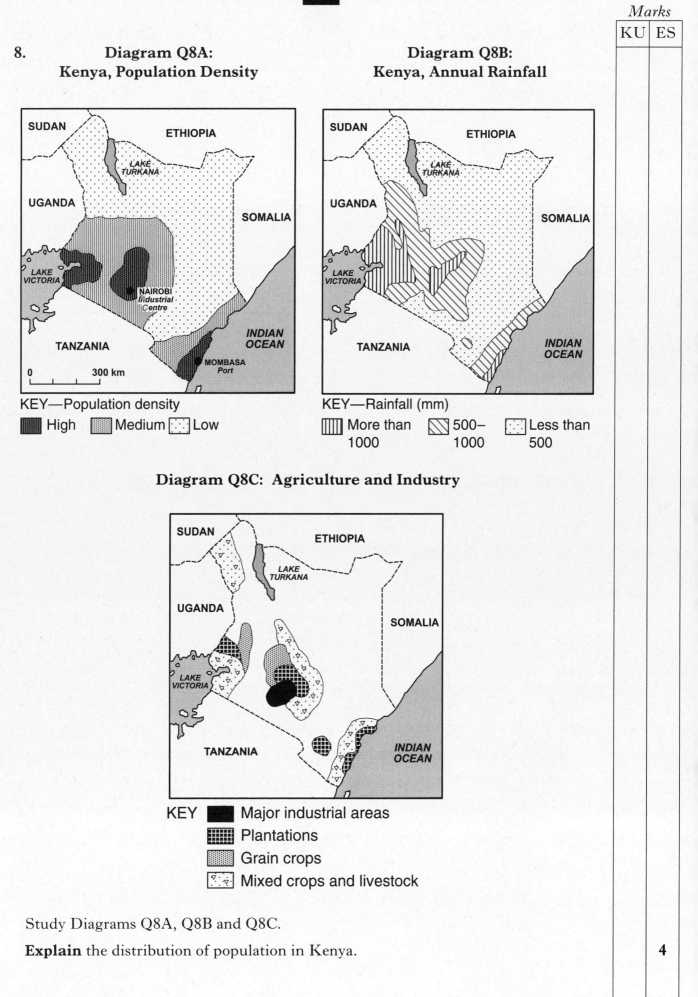

KEY—Population density

▉ High   ▨ Medium   ⬚ Low

KEY—Rainfall (mm)

▥ More than
1000

▧ 500–
1000

⬚ Less than
500

Diagram Q8C: Agriculture and Industry

KEY   ■ Major industrial areas

▦ Plantations

⬚ Grain crops

▽ Mixed crops and livestock

Study Diagrams Q8A, Q8B and Q8C.

Explain the distribution of population in Kenya.

4

Marks
KU | ES

9.

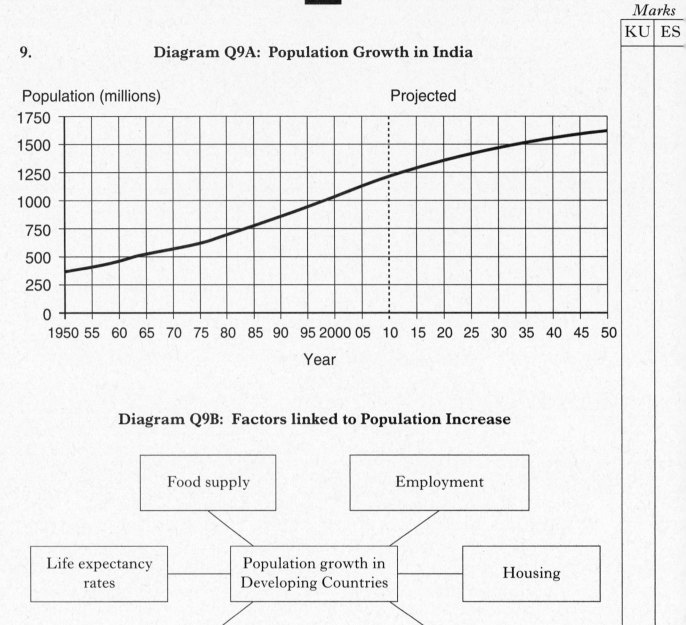

Diagram Q9A: Population Growth in India

Diagram Q9B: Factors linked to Population Increase

Look at Diagrams Q9A and Q9B.

"The population growth of India will cause the country many problems."

Do you agree fully with this statement?

Give reasons for your answer.

5

Marks
KU | ES

10.   **Diagram Q10: Selected Developing Countries where one raw material is more than half of all exports**

Look at Diagram Q10.

(*a*) What problems does this pattern of trade cause in developing countries?   **4**

(*b*) Which **other** processing techniques could be used to display the export figures shown on Diagram Q10?

Give reasons for your choice of techniques.   **5**

[Turn over for Question 11 on *Page twelve*

11. **Diagram Q11: Some Charities providing Aid to Developing Countries**

Look at Diagram Q11.

In what ways is aid given by charities (voluntary aid) suitable for developing countries?

4

[END OF QUESTION PAPER]

[BLANK PAGE]

[BLANK PAGE]

[BLANK PAGE]

Acknowledgements

Permission has been sought from all relevant copyright holders and Bright Red Publishing is grateful for the use of the following:

The European Emblem © The Council of Europe (2009 page 15);

Logo © Action Aid (2010 Credit page 12);

Logo © World Vision (2010 Credit page 12);

Logo © Water Aid (2010 Credit page 12);

The Oxfam Logo, is reproduced with the permission of Oxfam GB, Oxfam House, John Smith Drive, Cowley, Oxford OX4 2JY, UK www.oxfam.org.uk. Oxfam GB does not necessarily endorse any text or activities that accompany the materials. (2010 Credit page 12);

Ordnance Survey © Crown Copyright. All rights reserved. Licence number 100049324.